Essential Laboratory Skills for Biosciences

Essential Laboratory Skills for Biosciences

M.S. Meah and E. Kebede-Westhead
University of East London

A John Wiley & Sons, Ltd., Publication

This edition first published 2012 © 2012 by John Wiley & Sons, Ltd.

Wiley-Blackwell is an imprint of John Wiley & Sons, formed by the merger of Wiley's global Scientific, Technical and Medical business with Blackwell Publishing.

Registered office: John Wiley & Sons, Ltd, The Atrium, Southern Gate, Chichester, West Sussex, PO19 8SQ, UK

Editorial offices: 9600 Garsington Road, Oxford, OX4 2DQ, UK
                       The Atrium, Southern Gate, Chichester, West Sussex, PO19 8SQ, UK
                       111 River Street, Hoboken, NJ 07030-5774, USA

For details of our global editorial offices, for customer services and for information about how to apply for permission to reuse the copyright material in this book please see our website at www.wiley.com/wiley-blackwell

The right of the author to be identified as the author of this work has been asserted in accordance with the UK Copyright, Designs and Patents Act 1988.

Library of Congress Cataloging-in-Publication Data

Meah, Mohammed.
    Essential laboratory skills for biosciences / Mohammed Meah and Elizabeth Kebede-Weshead.
       p. cm.
    Includes bibliographical references and index.
    ISBN 978-0-470-68647-8 (pbk.)
1.   Chemistry–Laboratory manuals. 2.   Chemical apparatus–Handbooks, manuals, etc.
I. Kebede-Weshead, Elizabeth. II. Title.
    QD45.M39 2011
    507.2′1 – dc23

                                                                                2011043893

A catalogue record for this book is available from the British Library.

Wiley also publishes its books in a variety of electronic formats. Some content that appears in print may not be available in electronic books.

Typeset in 9/10.5 pt Times by Laserwords Private Limited, Chennai, India
Printed and bound in Malaysia by Vivar Printing Sdn Bhd

4 2014

Contents

List of Figures

List of Tables

Acknowledgements

We are extremely grateful to Varoopah Senthuran for taking the majority of photographs, and her enthusiasm, support and assistance in a variety of other tasks during the production of this book. Discussions on book contents with John Allum and Alberto Sanchez-Medina at the initial stage of writing the book were highly appreciated by the authors. We are also grateful to Stefano Casalotti and Winston Morgan for their suggestions and helpful feedback on the manuscript. We would also like to thank the following, for their help in providing the various apparatus in different laboratories, and for suggestions in taking better photographs: Susan Harrison, Raymond Stoker, Stephen Garrad, Duncan Kenedy, Kevin Clough, Manchu Ambihaiphan, Ashford Clovis, and Keith Eley.

We would also like to thank the publishing editors of Wiley Publishers for their patience, understanding and guidance in the preparation of this book.

Introduction

'Biosciences' is a broad term covering a wide range of subjects, including biology, biochemistry, biomedical science, biotechnology, forensic science, microbiology, physiology, pharmacology, and toxicology. Studying the biosciences involves a large amount of practical work in laboratories. These sessions involve following instructions from the lab instructor, following a practical schedule, learning techniques, taking measurements, observing and recording data, calculating and presenting data.

You will find that in performing practical laboratory tasks, techniques are often repeated (e.g. pipetting volumes, preparing solutions or producing calibration curves). You will be required to apply these fundamental skills across the bioscience subject areas.

This book is intended to act as a laboratory guide to help you to complete these practical tasks successfully by focusing on:

- the essential steps in learning a technique or using specific equipment;
- carrying out necessary calculations;
- presenting your results appropriately.

The book is intended for those who are new to, or have little experience of the laboratory techniques and equipment used in studying biosciences. It may also be useful to those who need to refresh some of their knowledge of laboratory skills, or those doing projects at both undergraduate and postgraduate levels. This book emphasizes 'essential' skills and the practical steps required to use equipment and learn techniques. The authors assume that the reader has very little prior knowledge of techniques in the biosciences. To make the process easier, we have emphasized the step-by-step approach in the practical procedures, together with a wide range of photographs of equipment and accessories.

This is not intended to be a comprehensive text book, so you will find that the background theory to the techniques (Chapters 3, 4, 5 and 6) is brief. Also, some techniques in the various bioscience fields, and some particularly advanced techniques that you may be using in years two and three of your degree may not have been included. Please note that the steps shown for procedures are not universal, and you will come across some differences in practice, depending on the technique and equipment you are using.

Chapter 1 covers the fundamental basis of measurements and units, and common mathematical calculations you may come across. You must understand terms which are commonly used, such as 'moles', 'molarity' and 'dilutions'. In addition, you must be able to convert between common prefixes or powers of ten. You will have to practise doing calculations involving moles, molarity and powers. Pictures of glassware commonly used in the lab are included.

Chapter 2 gives details of common terms associated with solutions and how different types of solutions are prepared.

Chapter 3 contains details of techniques used to separate liquids and solids, such as filtration, centrifugation, chromatography and gel electrophoresis. Some of the chromatography methods using expensive equipment will not be routinely used in undergraduate bioscience practicals, especially at entry level. Therefore, instead of detailed procedures of the methods, we give the basic principles and an outline of the methods. It is intended to help distinguish the methods which may appear confusing for a beginner.

Chapter 4 contains details of a wider variety of techniques and equipment, including spectrophotometry, pH meters, titration and aseptic techniques.

Chapter 5 contains details on light microscopy, histology, and cell counting.

Chapter 6 covers a range of techniques, investigating cardiopulmonary function in humans.

Chapter 7, the final chapter, describes how to record and present your data, with examples of typical tables and graphs.

Appendices in bold face to highlight the section give more details of laws of powers and logarithms, more theory on some techniques (e.g. spectrophotometry), and the use of common software such as Microsoft *Word* and *Excel* to produce tables, graphs and descriptive statistics. There is only a brief description of statistical analysis (descriptive statistics), so you will need to refer to statistical texts to investigate differences in mean data (inferential statistics).

We hope this book will be a useful 'companion' to the lab schedule and lecturers' instructions, and will help to make the student more independent in the laboratory.

Mohammed Meah and Elizabeth Kebede-Westhead
University of East London

Health and Safety

Before performing any laboratory work, it is essential that health and safety considerations are followed with regard to hazards and risk to you and others. Make sure you follow the basic important rules in a laboratory as shown below.

- Know the lab safety rules.
- Be aware of the COSHH (Control of Substances Hazardous to Health) forms which should be displayed in the laboratory.
- Notify the instructor of any medical condition or allergies.
- Know the locations of the closest fire extinguisher, fire alarm, eyewash and other emergency equipment, and know any fire regulations.
- Use appropriate protection, e.g. protective clothing, lab coats, gloves, goggles, fume cupboards.
- Familiarize yourself with how to use the equipment you are working with, and be aware of any precautions you have to take in its use.
- Always follow instructions and use equipment properly; if you are not sure, ask the instructor or technician.
- Do not return any substance back into original containers.
- Know how to dispose of waste (solutions, sharps, pipette tips, etc.).
- Report all injuries to your instructor/technician.
- Report any spills or breakages or injury to the instructor or technician.
- Keep your work area clean and return equipment.
- At the end of the laboratory session, wash your hands thoroughly.

1

Measurements and Calculations

Learning outcomes

- To know the common measurements and their units.
- To learn the factors which affect measurements.
- To recognize the main apparatus used in volume measurements.
- To recognize the different types of pipettes and how they are used to measure specific volumes.
- To use balances to weigh out a known weight of a substance.

1.1 Units and measurements

In the study of bioscience subjects, a wide range of measurements of variables are made using various kinds of equipment. Each measurement is described by a number and a unit.

The Systeme Internationale D'Unites (SI – also known as the *metric system*) is the international system of units used to describe data. Some common measurements are shown in Table 1.1.

Some measurements, such as pH or absorbance of light, have no units.

When we are dealing with very small or large values, prefixes are used for convenience. The most common ones are shown in Table 1.2.

All measurements are affected by errors, which can be random (cannot be predicted) or systematic (biased, predictable). The measurements made can be described by:

- **accuracy** – how close the values are to the true values, i.e. the average value will be nearer the true value but individual measurements may not necessarily be close to each other (see Figure 1.1); or
- **precision** – how close together are the spread of the values, i.e. repeated measurements will be similar but not necessarily close to the true value.

Essential Laboratory Skills for Biosciences, First Edition. M.S. Meah and E. Kebede-Westhead.
© 2012 John Wiley & Sons, Ltd. Published 2012 by John Wiley & Sons, Ltd.

Table 1.1 *Common measurements and their units and symbols*

Measurements	Units	Symbols
Length	metre, centimetre, millimetre	m, cm, mm
Mass	kilogram, gram	kg, g
Time	second, minute, hour	s, min, hr
Volume	cubic metre, litre, cubic decimetre, millilitre, cubic centimetre	m^3, l, dm^3, ml, cm^3
Concentration	molar, mole per litre, parts per million	M, $mol\ l^{-1}$, ppm
Amount of substance	mole	mol
Density	kilogram per cubic metre, gram per cubic centimetre	$kg\ m^{-3}$, $g\ cm^{-3}$
Force	Newton	N
Pressure	Newton per square metre, Pascal, millimetres of mercury	$N\ m^{-2}$, Pa, mm Hg
Energy	Joule	J

Table 1.2 *Common prefixes*

Power	Prefix	Symbol
10^{-9}	nano	n
10^{-6}	micro	μ
10^{-3}	milli	m
10^{3}	kilo	k
10^{6}	mega	M
10^{9}	giga	G

Figure 1.1 shows the results of firing a pellet gun eight times on each of four targets. This illustrates the combinations of accuracy and precision obtained by repeated firing at the targets.

Errors in the accuracy and precision can be calculated (see Appendix 5).

To reduce error, measurements are repeated (replicated). They may also be repeated at another time to see if the same results are reproducible. One important factor in ensuring accuracy of measurements is performing regular *calibration* of equipment – the process of checking the accuracy of equipment using known values and standards over the measurement range under standard conditions.

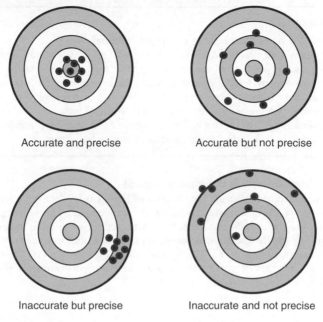

Accurate and precise

Accurate but not precise

Inaccurate but precise

Inaccurate and not precise

Figure 1.1 *Accuracy and precision shown on a target*

It must not be assumed that all equipment has been calibrated prior to use. It is common practice that all items of equipment are calibrated against a known standard and are checked regularly to ensure that measurements are reliable. For example, pH meters are calibrated with buffers of known pH, commonly with pH 4 and 7. Depending on the type of equipment and ease of calibration, it is possible for students to perform simple calibration in lab sessions.

Example 1.1

Check a weighing balance by using known measurements of volume.
Weigh a beaker repeatedly by adding a series of 1 ml of water. Assuming the density of water is about $1.00 \, g \, ml^{-1}$, the weight should increase by 1 g with every addition of 1 ml.

Example 1.2

Check a pipette by weighing dispensed volume.

Using a 1 ml pipette, dispense aliquots of 1 ml of water into a small beaker and weigh the volume repeatedly after each addition. As the volume increases by 1 ml, the weight should increase by 1 g. You may need to pipette 5–10 ml, as the measurement may not be sensitive with small volumes if you are not using a sufficiently sensitive balance.

Calibration is also performed in another context, namely to determine unknown concentrations using standard solutions (see Chapter 4).

1.2 Measuring the volumes of liquids

The first step in measuring the volumes of liquids is to identify the glassware or equipment that would be most appropriate to use. This will depend on the volume to be measured and the accuracy required in the measurement. The smaller the amount you need to measure, the more sensitive the equipment will need to be. The figures below (Figures 1.2a–2f) illustrate the various types of volumetric labware that you will commonly use in the lab.

Beaker

- Graduated

- Rough measure

- Used for routine dispensing, mixing solutions, not for accurate measurements

- Volume labels are approximate, accurate to within 5%

Figure 1.2a Beaker

Conical (Erlenmeyer) flask

- Graduated

- Rough measure

- Used for routine dispensing, mixing solutions, not for accurate measurements

- Volume labels are approximate, accurate to within 5%

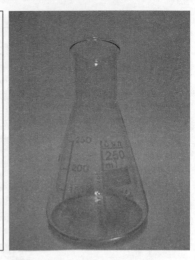

Figure 1.2b Conical (Erlenmeyer) flask

Measuring Cylinder

- Also known as graduated cylinder

- For general purpose use in measuring liquid volumes

- More accurate than beakers or flasks

- Accurate to within 1%

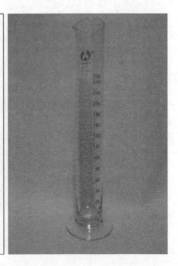

Figure 1.2c Measuring cylinder

Burette

- Graduated, with a stopcock at the bottom to dispense liquid precisely

- Used mainly for titration to deliver volumes accurately

- Funnel is used to fill the burette, and readings are taken at the bottom of the meniscus, starting at 0 from the top

- A stopcock is used to deliver the solution

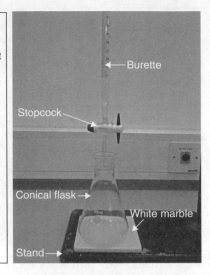

Figure 1.2d *Burette*

Volumetric Flask

- Calibrated to deliver only one specific volume; e.g. 25, 50, 1000 ml

- Used to prepare stock and standard solutions very accurately

- Narrow neck with etched ring which indicates the volume, e.g. 500 ml ± 0.2 ml.

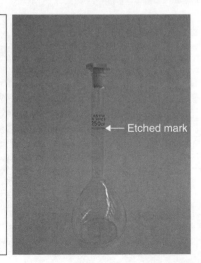

Figure 1.2e *Volumetric flask*

Pipette

- Used to measure and transfer small amounts of solution accurately

- Device with a mechanism for drawing solution into the pipette and for dispensing (e.g. bulb, plunger)

- Several design variations

- Common lab use: Gilson-type pipettes - various size ranges; different volumes shown here are P20, P200, P1000, P5000.

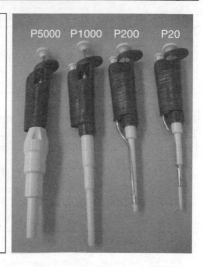

Figure 1.2f *Pipette*

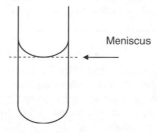

Figure 1.3 *Location of the meniscus of a liquid*

1.2.1 Reading at the meniscus

The surface of a liquid is curved rather than flat, and it is called the *meniscus* (see Figure 1.3). This curvature is due to the relatively strong attractive force between the glass and water molecules (surface tension); it is more obvious in narrow containers.

When reading a volume, precise reading should be taken at the lowest point of the meniscus, which is determined by looking at the liquid surface at eye level. It should not be read from above or below, as this would lead to an error called parallax. If the eyes are below the meniscus level, then the value will be lower than the correct volume; conversely, if the eyes are above the meniscus, then the volume will seem to be higher.

Table 1.3 *Common measurement units of volume*

	Large	→	Small
Units	litre (l)	millilitre (ml)	microlitre (μl)
Equivalents	1 l	1000 ml	1 000 000 μl
		10^3 ml	10^6 μl

1.2.2 Common units of volume measurement

The standard unit of volume used in the metric system is the litre (l) and its variations, the millilitre (ml) and microlitre (μl), etc.

To convert from a larger to a smaller measurement unit (Table 1.3), we would multiply (e.g. by 1000 from l to ml, by 1000 from ml to μl).

To convert from a smaller to larger measurement unit, we would divide (e.g. by 1000 from μl to ml, and by 1000 from ml to l).

Example 1.3

a) Express 0.85 l in ml → $0.85 \times 10^3 = 850$ ml
b) Express 850 μl in ml → $850 \div 10^3 = 0.850$ ml
c) Express 1.5 ml in μl → $1.5 \times 10^3 = 1500$ μl

1.3 Pipetting

Pipettes are used to measure and dispense accurate volumes of solutions. There are various different types of pipettes in use in laboratories. Although most of these are similar, their settings for adjusting the volume differ. Three popular types of pipettes which are commonly used in the laboratory are described below.

1.3.1 Gilson Pipetman

These are adjustable mechanical pipettes that come in various sizes (Table 1.4 and Figure 1.5), and the most frequently used sizes lie in the range between 0.2 μl to 10 ml. They are autopipettors which work by air displacement. They consist of the following parts (see Figure 1.4):

(i) a push button plunger (to uptake and dispense liquid)
(ii) adjustment ring and volume scale (to set the correct volume)
(iii) a barrel to which a disposable tip is attached
(iv) tip ejector button and tip ejector (to expel the tip after dispensing liquid).

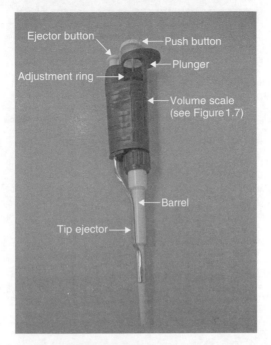

Figure 1.4 *Parts of a Gilson pipette*

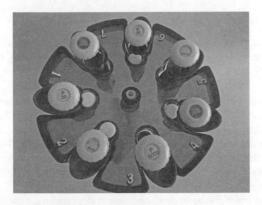

Figure 1.5 *Circle of pipettes*

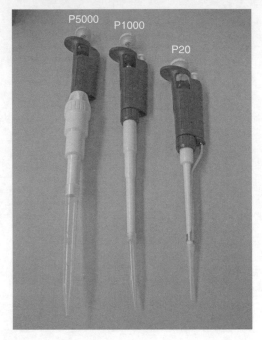

Figure 1.6 *Tips used with pipettes*

Note that there are also other types of pipettes very similar to the Gilson Pipetman (e.g. Eppendorf), but with slight variations in volume adjustments, such as on the adjustment rings or dials.

The P20 and P200 use yellow pipette tips, and the P1000 uses blue tips (Figure 1.6). The volume of each of the Gilson pipettes is set by turning the adjustment knob to get the desired volume in the window.

The reading window for the volume shows three levels (Figure 1.7), each level denoting a difference of one order of magnitude. For example, with a P1000 pipette, a dial reading 1–0–0 is set to pipette 1000 µl, while a dial reading 0–1–0 is set to pipette 100 µl.

Choosing the right pipette is important if the required volume is to be measured accurately. For example, although a 10 ml pipette has a range of 1–10 ml, a 5 ml pipette should be used to pipette accurately 5 ml or

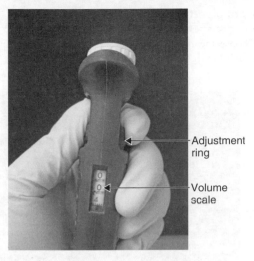

Figure 1.7 Setting the volume on a Gilson pipette

less. Similarly, a P200 with the range of 50–200 µl can deliver 100 µl, but it is more accurate to use a P100.

Using the wrong pipette is frequently a source of common error in practical work involving accurate volume measurements, such as preparing standard solutions. First note the volume required, then choose the pipette which has the closest range to this (see Table 1.4).

Table 1.4 Volume ranges for Gilson pipettes

Pipette	Minimum recommended volume	Maximum volume
P2	0.5 µl	2 µl
P10	2	10
P20	5	20
P100	20	100
P200	50	200
P1000	200	1000 µl = 1 ml
P5000	1 ml	5 ml
P10 000	1 ml	10 ml

1.3.2 How to pipette

Step 1: Choose the right pipette for the volume range required and attach the right coloured tip (Figure 1.8).

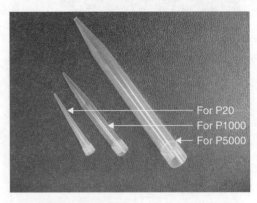

For P20
For P1000
For P5000

Figure 1.8 *Types of pipette tips*

Figure 1.9 *Pipette inserted into beaker*

Step 2: Set the volume to be measured by turning the dial. Note there are only three spaces for setting the volume in the window. For example, for the P20, in order to set 5, 10, 15 and 20 µl, you would set the dial (from the top) to 0-5-0, 1-0-0, 1-5-0 and 2-0-0 respectively.

Step 3: Practise pushing the plunger to find the two 'stop' positions (opposition or resistance to pushing plunger).

Step 4: Push the plunger to the first 'stop', then place the tip of the pipette into the solution and draw the solution into the pipette tip (Figure 1.9).

Step 5: Place the pipette into the receiving tube or container and push the plunger all the way to the second 'stop' position to ensure that all the solution is pushed out of the pipette tip.

Step 6: Eject the tip into waste using the ejector behind the plunger (Figure 1.10).

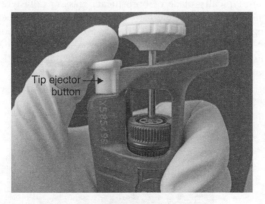

Figure 1.10 *Ejecting tip using tip ejector button*

 Precautions in handling pipettes

- Don't forget to use the pipette tip and make sure it fits correctly.
- Don't force the plunger.
- Always put in a pipette holder when not in use (look at Figure 1.5).
- Don't walk around with a pipette.

- Hold the pipette upright, and do not leave it on bench at an angle as its contents may flow backwards, losing liquid and blocking the pipette.
- Inspect the tip for blockages or dirt.
- Never use 5 ml or 19 ml pipette without a filter.

1.3.3 Multichannel pipettes

For repetitive tasks (for example, filling a 96-well tray), a multichannel pipette is used. These draw and expel the same volume in each of the individual tubes making up the multichannel pipette (see Figure 1.11).

1.3.4 Vacuum pipettes

Different from the air displacement pipettes described above (1.3.1–1.3.3), in these pipettes where liquid is drawn up a cylinder by filling up a vacuum which is created by an attached pipette filler, pipette pumps or bulbs. The pipettes are typically made of borosilicate glass or polystyrene, and can be autoclavable or disposable for use in sterile work. Volumetric pipettes dispensing single volumes (e.g. 10, 25, 50 ml) are commonly used in preparing large quantities or stock solutions. Variable quantities of liquids can be dispensed with pipettes that are graduated along the length of the cylinders.

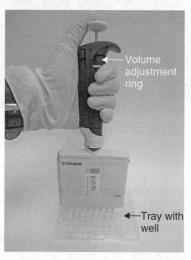

Figure 1.11 *Multichannel pipette*

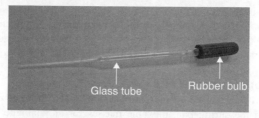

Figure 1.12 *Pasteur pipette*

Pasteur pipette (Figure 1.12) is another type of pipette commonly used in the lab for transferring or dispensing small amounts of liquid. These are small glass tubes with an attached rubber bulb which can be squeezed to create a vacuum and liquid is sucked up when released, but it is not quantitative.

1.4 Weighing

You will be frequently asked to weigh small amounts of chemicals to make up solutions. This will involve using balances (Figure 1.13), which will give the mass to either two or three decimal places in grams. A common

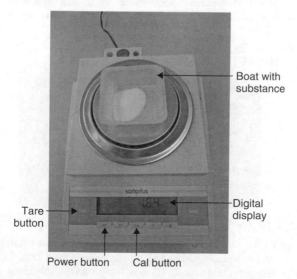

Figure 1.13 *Portable digital top-loading balance*

procedure in the use of these balances is *taring*, which means to remove the weight of the container holding the chemical by setting the scale to zero while the container is on the balance. Increases in the reading on the balance would then be accounted to the mass of the substance added.

1.4.1 Top-loading balance

A typical balance used to measure mass will consist of a zero button or knob, a measuring pan, movable masses for an analogue balance or a power button for a digital balance. A portable top-loading balance (Figure 1.13) is most frequently used for routine laboratory procedures such as making up solutions. It can show accuracy to three or four decimal places.

Measurement range is usually 0.01–0.1 g or 1–50 g.

How to weigh using a top-loading balance

> **Step 1:** Place the balance on a clean, level bench away from draughts and vibrations.
> **Step 2:** Check that the balance pan is clean.
> **Step 3:** Check the balance is level – the spirit level should be in the centre. If not, use the adjustable feet to make it level.
> **Step 4:** Put an empty beaker, a boat (Figure 1.14) or foil on the pan.
> **Step 5:** Press the Tare button. Always press this before any new measurement.
>
>
>
> *Figure 1.14 Plastic boat*
>
> **Step 6:** Check that the scale displays zero and the correct number of decimal places, depending on the balance used (e.g. 0.000).
> **Step 7:** Using a clean spatula (Figure 1.15), add the substance to the centre of the container.

Step 8: Keep adding the substance, reducing the amount added when approaching close to the target mass.

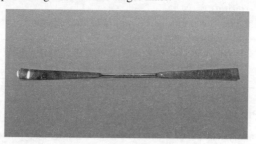

Figure 1.15 Spatula

Step 9: If you go over the target mass, carefully remove small amounts and discard safely. Do not return the substance back into the container.

Step 10: Record the mass when the display shows a constant reading.

 Precautions

Any spills should be cleaned with a brush and disposed carefully, but make sure the balance is switched off before cleaning.

1.4.2 Analytical balance

A more sensitive balance used for weighing very low masses (0.0001 g) is the analytical balance (see Figure 1.16). This is a high-precision beam balance composed of a level with two equal arms and a pan suspended from each arm. It is enclosed by glass, with two sliding sides. Measurement range is typically 0.01–0.10 mg.

How to weigh using an analytical balance

Step 1: Switch the balance on and the display should read zero (0.0000).

Step 2: Place a small weighing boat on the weighing pan.

17

Step 3: Close the glass doors and wait a few seconds.

Step 4: Press the control button so that the display again reads 0.0000 (taring).

Step 5: Open the doors gently and add the substance carefully until the required mass is reached.

Step 6: Again close the doors and wait until the reading is stable, then record the final mass.

Precautions

- An analytical balance is a sensitive instrument, so make sure you do not make any sudden movements that will shift the balance.
- Any spill should be cleaned carefully before proceeding with more measurements, as it could be a danger to the next person who works in this area, who will not know what the chemical left behind is – an unknown white powder could be just salt or it might be cyanide!

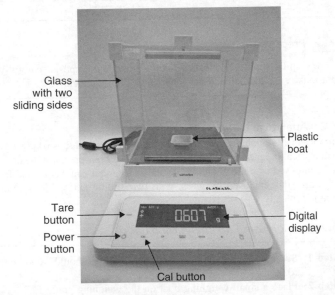

Figure 1.16 A typical analytical balance

Conversions between units of volume and mass

Volume

$$1\,l = 1000\,ml$$
$$1\,ml = 1000\,\mu l$$

Mass

$$1\,kg = 1000\,g$$
$$1\,g = 1000\,mg$$
$$1\,mg = 1000\,\mu g$$
$$1\,\mu g = 1000\,ng$$
$$1\,ng = 1000\,pg$$

1.5 Calculations

All quantities of measurements need to be reported by a *number* and the *unit* of measurement, except in cases such as absorbance readings on a spectrophotometer (in which case the wavelength at which absorbance was read should be included).

When using units, the numbers should be separated from the unit by a space. The forward slash means 'per' or 'divide by' – for example, 5 mg/l is read as 'five milligrams per litre'. The slash is replaced by a superscripted $^{-1}$ after the unit in scientific writings and publications; so mg/l should correctly be written as $mg\,l^{-1}$.

1.5.1 Significant figures

Significant figures provide a way to express the degree of accuracy in experimental data which should be recorded with as many digits as can be accurately measured. The number of significant figures is the number of non-zero values in a measurement. Where there are many calculations on the same data, then the final value would be rounded up (i.e. if >5 then increase by one; if <5 then leave as it is).

Example 1.4

a) 7849 to 2 significant figures = 7800
b) 0.0457 to 2 significant figures = 0.046
c) 0.0457 to 2 decimal places = 0.05

1.5.2 Powers

In the decimal counting system that we use, the values of numbers are represented by columns which represent multiples or divisions of 10. For example, the number 1451.25 represents 1 thousand, 4 hundreds, 5 tens, 1 unit, 2 tenths and 5 hundredths. The value of these numbers are so because the columns to the left of the decimal point are represented by multiples of 10, i.e. units (10^0), tens (10^1), hundreds (10^2), thousands (10^3) and so on, while the columns to the right of the decimal point are represented by divisions of 10, i.e. tenths (10^{-1}), hundredths (10^{-2}), thousandths (10^{-3}), etc. This factor of 10 by which the number is increasing or decreasing is called the *base*, and the superscript is the *power* or *exponent*. For example, 10^{-2} has the base as 10 and the power as $^{-2}$.

In general, when a number is multiplied by itself a number of times or divided by itself a number of times, the number of times is called the *power* and the number is called the *base*. For example, $10 \times 10 \times 10$, can also be written as 10^3, where 10 is the base and the superscripted 3 is the power.

Similarly, three divisions of 10 ($10 \div 10 \div 10$ or $\frac{1}{10} \times \frac{1}{10} \times \frac{1}{10}$) can be written as 10^{-3}, where 10 is the base and -3 is the power. Other bases can also be used, the best known being the binary system used in computers, which has base 2 – i.e. $2^0, 2^1$, etc.

Powers are very useful in dealing with very large or small numbers, as in the table of prefixes elsewhere in this chapter (Table 1.2, page 2). The rules for calculations involving powers are shown in Appendix 1.

Converting between units involving powers of 10

It is very useful to be able to quickly convert between units, particularly between prefixes changing by a constant factor such as 1000, for example in volume units (l, ml and μl) and weight units (g, mg and μg).

To convert from a large unit to a small unit, multiply by 1000 for one prefix (e.g. 1l = 1000 ml) or 1000×1000 for two prefixes (e.g. 1l = 1000×1000 μl).

To convert from a small unit to a larger unit, divide by 1000 for one prefix (e.g. 1 μg = $\frac{1}{1000}$ mg) or divide by 1000×1000 for two prefixes (e.g. 1 μg = $\frac{1}{1000} \times \frac{1}{1000}$ g).

1.5.3 Logarithms

One application of powers is in the use of logarithms or logs, which is useful in dealing with large or small changes in values. Examples include the pH scale (Chapter 3), exponential bacterial cell growth over time, or dose response curves in pharmacology.

In such cases, it is common to use *scientific notations* – the expression of numbers using multiples (prefixes) of base 10.

e.g.

$$0.0052 = 5.2 \times 10^{-3}; 0.003 = 3 \times 10^{-3}$$

The logarithm is *the power to which a base number must be raised to give that number*. For example, what power of 10 will give 1000? The answer is 3, since 10^3 is equal to 1000. Logs to base 10 (simply called 'logs') are important because as a number goes up by a factor of 10 (i.e., multiplied by 10), the log goes up by one.

The two most commonly used base numbers are 10, where the power is $\log_{10}$, and the number 2.714 (also known as 'e'). Logs to base 'e' are called natural logs and are written as $\log_e$ or simply 'ln'. Logarithms do not have units.

Rules for calculations involving logs are shown in Appendix 2.

Example 1.5

a) 10^1 (or 10) $\log 10^1$ $= 1$
b) 10^2 (or 100) $\log 10^2 = 2$
c) 10^3 (or 1000) $\log 10^3 = 3$
d) 10^{-2} (or 0.01) $\log 10^{-2} = -2$

1.5.4 Fractions, ratios and percentages

A *fraction* is the division of one number (numerator) by another (denominator), with the answer being either less than 1 (e.g. $\frac{1}{2}$) or greater than 1 (e.g., $\frac{7}{5}$). In these two examples, 1 and 7 represent the numerators and 2 and 5 represent the denominators.

A *ratio* is simply a different way of expressing a fraction. Ratios are used to express proportions or fractions, usually in integers (e.g., the fraction $\frac{1}{2}$ expressed as a ratio would be written 1:2 and expressed as 'a ratio of 1 to 2'. Ratios do not have units).

A *percentage* is a fraction in which the denominator is always 100. The symbol used for percentage is '%'. Thus 20% is equal to $\frac{20}{100}$ and is also 0.20.

Example 1.6

Find the ratio of three substances in a mixture with the quantities 20 g:15 g:30 g.

Divide by the highest common factor (HCF) which will go exactly into each of the quantities; in this case, the HCF would be 5, giving the ratio 4:3:6.

Ratios and dilutions

Ratios are used frequently in the process of making dilutions of liquids. If you are asked to do dilutions in parts or ratios, this specifies the fractions to combine (see Chapter 2). If asked to dilute to a dilution factor, you need to calculate how many times the final volume must be increased (see Chapter 5 – Cell counting).

Example 1.7

a) Perform a 1 in 10 (1:10) dilution.
   Take 1 ml of substance and add 9 ml of water or buffer.
b) Dilute 20 μl of NaCl stock solution 200-fold to a final volume of 4 ml (or 4000 μl) using distilled water.
   Multiply 200 by 20 (= 4000); but you need to subtract the volume of the stock solution that you are using from the total volume. Therefore, the volume of water to be added is

$$4000 - 20\,\mu l = 3980\,\mu l.$$

2

Preparing Solutions

Learning outcomes

- Know common terms associated with solutions.
- Be able to make up different solutions.
- Be able to do common calculations involving solutions.

2.1 Common terms defining solutions

In a solution which is a simple mixture of two or more substances, the substance dissolved in solution is called the *solute*. The substance the solute is dissolved in is called the *solvent*. For example, if you dissolve sugar in water, the sugar is the solute and the water is the solvent. In other words, a solution is made by dissolving a solute in a solvent.

2.1.1 Stock solution

A *stock solution* is one which has a known volume and concentration and is usually in a *concentrated* form. It is often used to make a range of solutions of different concentrations by *dilution* (spreading a given amount of solute over a larger solution volume).

2.1.2 Standard solutions

Various *standard solutions* are made up from a stock solution. Their volumes are constant but their concentrations vary up to the concentration of the stock solution. They are produced from the stock solution by dilution and they are used in drawing a calibration or standard curve (see Chapter 4).

2.2 Precautions in making solutions

2.2.1 Solute

Check the label on the chemical bottle container and note the formula and molecular weight. Be aware of properties such as solubility, cost and

---

Essential Laboratory Skills for Biosciences, First Edition. M.S. Meah and E. Kebede-Westhead.
© 2012 John Wiley & Sons, Ltd. Published 2012 by John Wiley & Sons, Ltd.

Figure 2.1 Wash bottle

stability. Check that the substance is not out of date, nor discoloured or decomposed. Check if the substance is hazardous and take appropriate precautions (e.g. use gloves, fume cupboard, etc.). For further details on the chemicals, go to the manufacturer's website.

2.2.2 Solvent

Never use tap water, as it contains impurities. Use distilled or deionized water, as usually dispensed from large laboratory containers. Do not use distilled water from 'wash bottles' or plastic bottles with a spout (Figure 2.1) for making solutions, as this may contain impurities. Wash bottles are used mainly to rinse glass and plastic labware.

Other factors to note when making solutions include temperature, solubility, type of solute and type of solution (see Appendix 3).

2.3 Making solutions

2.3.1 Making an aqueous solution

> **Step 1:** Weigh the required amount of solute, using a pan or analytical balance (see Chapter 1).
>
> **Step 2:** Add this to a volumetric flask (from a weighing boat or aluminium foil) by tilting the volumetric flask.

Step 3: Add a little volume of distilled water from a beaker, washing down solids left behind near the neck, and then add the stopper and mix by shaking the flask carefully. Make sure the stopper is secure before shaking.

Step 4: Place the flask on the bench and add more distilled water until you are 1–2 cm below the etched mark.

Step 5: The final filling should be done drop by drop, using a Pasteur pipette.

Step 6: Use the concave meniscus for accuracy and to avoid parallax error, and see that the bottom curve of the meniscus is on the etched line (see Chapter 1).

Step 7: Add the stopper and do a final mix.

2.3.2 Making diluted solutions

A common procedure used in bioscience practical work is to produce a range of diluted solutions from a more concentrated solution or a stock solution. The two most common types of dilutions in the lab are serial dilutions and dilutions for preparing a range of standard solutions. If an error is made, it only affects one dilution in the latter case.

2.3.3 Serial dilutions

Serial dilutions is the term given to a series of dilutions prepared from a more concentrated solution. A common use of serial dilutions is to create a set of dilutions in which each one is one-tenth or one-half the concentration of the previous level. This is a common method in experimental biosciences, particularly involving logarithmic or semi-logarithmic responses, such as bacterial culture or media in microbiology, or in bioassays.

In the diagram shown below (Figure 2.2), the stock solution is serially diluted by a factor of ten for each tube, i.e.:

- 10 times by adding 1 ml of stock to 9 ml of water;
- 100 times by taking 1 ml from the 10× diluted tube and adding 9 ml of water;
- 1000 times by taking 1 ml from the 100× diluted tube and adding 9 ml of water;
- 10 000 times by taking 1 ml from the 1000× diluted tube and adding 9 ml of water.

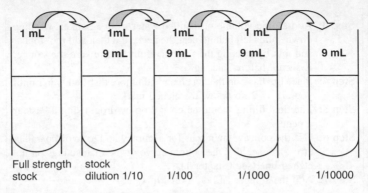

Figure 2.2 Serial dilution of a stock solution

If the concentration of the stock solution was $1\,mg\,ml^{-1}$, the series will have concentrations of $1 \times 10^{-1}, 1 \times 10^{-2}, 1 \times 10^{-3}, 1 \times 10^{-4}\,mg\,ml^{-1}$.

The advantage of this method is the ease of pipetting. The disadvantage is that an error in one dilution causes errors for all the subsequent dilutions.

2.4 Dilutions to prepare standard solutions

You will frequently be asked to make a series of standard solutions from a stock solution (see Example 2.2 below).

2.4.1 Calculations in dilution

A simple way of calculating the concentrations of a solution after dilution is by using the equation $C_1 V_1 = C_2 V_2$, where:

C_1 = concentration of stock solution
V_1 = volume of stock solution needed to make the new solution
C_2 = final concentration of new solution
V_2 = final volume of new solution

The principle is:

Number of moles of solute before dilution = number of moles of solute
after dilution.

No matter how much your sample is diluted, the total amount of your substance remains the same. The less the substance, the more the water, i.e. molarity × (volume before dilution) = molarity × (volume after dilution) – or $C_1V_1 = C_2V_2$.

Example 2.1

Given a stock solution = 0.1 M, prepare a new solution of 2 ml volume of 0.02 M concentration. How much volume of stock would you need?

$$C_1 = 0.1\,M;\ V_1 = ?;\ C_2 = 0.02\,M;\ V_2 = 2\,ml$$

$$\text{or } 0.1\,M \times V_1 = 0.02\,M \times 2\,ml$$

$$\text{or } V_1 = (0.02\,M \times 2\,ml)/0.1\,M$$

$$V_1 = 0.4\,ml$$

Example 2.2 Step-wise worked example

Using a stock solution of 1.5 mM of a substance, make a series of standard solutions of volume 3 ml.

Step 1: First make a table of the standard solutions you will need to make, e.g. six standard solutions: 0.0, 0.5, 0.75, 1.0, 1.25, 1.5 mM.

Step 2: Use the formula $C_1V_1 = C_2V_2$ to calculate the volume of stock needed. For example, to find the volume of stock to make 0.5 mM:

$$1.5 \times V_1 = 0.5 \times 3$$

$$\text{or } V_1 = 0.5 \times 3/1.5 = 1\,ml$$

Step 3: Pipette 1 ml of stock into a glass tube and add 2 ml distilled water to make up a total volume of 3 ml. Then label it clearly. Make sure labels are meaningful – not just 1, 2, 3, etc. – unless you write in your lab book what 1, 2, 3, stand for.

Step 4: Repeat steps 2 and 3 until all of the six standard solutions are made as shown in Table 2.1.

An alternative method of calculating the 0.5 mM standard solution in the above example would be:

Since the original concentration is 1.5 mM and the final concentration is 0.5 mM, the dilution factor is $0.5/1.5 = 0.33$ or a 1 in 3 dilution. Thus the amount of stock solution required is $\frac{1}{3}$ of 3 ml = 1 ml, and the volume of water needed is $3 - 1 = 2$ ml.

27

PREPARING SOLUTIONS

Table 2.1 *Standard solutions prepared from a stock solution by dilution*

Tube	1	2	3	4	5	6
Concentration (mM)	0.0	0.5	0.75	1.0	1.25	1.5
Volume of stock (ml)	0.0	1.0	1.5	2.0	2.5	3.0
Volume of distilled water (ml)	3.0	2.0	1.5	1.0	0.5	0.0

You can also cross-check your calculations by checking proportions, since the total volume is the same for all. For example, Tube 4 (1.0 mM) should have double the amount of stock solution as Tube 2 (0.5 mM) and half the amount of water.

2.4.2 Amount of a substance

Mole is a unit of measurement used to quantify the *amount* of a substance. The chemical definition of a mole is the amount of substance which contains as many particles as there are atoms in 12 g of carbon.

Example 2.3

1 mol of sodium (Na) has a mass of 23 g – the relative atomic mass (RAM) of Na.

1 mol of sodium chloride (NaCl) has a mass of 58.5 g: the relative molar mass of NaCl, i.e. add the atomic mass of Na (23) and chlorine (Cl = 35.5).

In simple terms, one mole of an element equals its atomic mass in grams. One mole of a molecule (two or more atoms bound together) is its molecular mass in grams; in other words, it is the total of the atomic masses of the elements in the molecule.

$$\text{Mole} = \text{Mass/Relative Atomic Mass (RAM)}$$

2.4.3 Concentration of a substance

Concentration is the term used to describe the strength of a solute in a solution. In other words, it is a measure of the amount of a substance dissolved in a given volume of liquid.

$$\text{Concentration} = \text{amount}/\text{volume}$$

Molarity is a term used to define concentration, and molar (M) is the unit of concentration. A one molar (1 M) solution contains one mol of dissolved substance in 1 litre.

$$\text{Molarity (M)} = \text{amount of substance (mol)}/\text{volume (l)}$$

2.5 Molar solutions

A 1.0 Molar (1.0 M) solution is equivalent to 1 formula weight (g mole^{-1}) of chemical dissolved in water to make 1 L. The formula weight (FW) is given on the label of a chemical bottle, or molecular weight (MW) can be used if formula weight is not given.

- A 1 M solution contains 1 mol of substance dissolved in 1 L solution
- 1 mol is equal to the molecular weight or atomic weight of the solute in grams
- 1 mol = 1000 m mol
- 1 mmol = 1000 µmol

Example 2.4

1 M solution of NaCl contains 58.5 g Na Cl dissolved in 1 l of solution.

$$1 \text{ mol in } 1 \text{ l} = 1 \text{ M}$$
$$0.5 \text{ mol in } 0.5 \text{ l} = 1 \text{ M}$$
$$0.1 \text{ mol in } 1 \text{ l} = 0.1 \text{ M}$$

Example 2.5

A solution of 1 M glucose contains 180 g (1 mol) dissolved in 1 l (1000 ml), which also means 18 g dissolved in 100 ml = 18%.

A solution of 1 M sodium chloride (NaCl) contains 58.5 g (1 mol) dissolved in 1 l (1000 ml), which also means 5.85 g in 100 ml = 5.85%.

Example 2.6

What is the molar concentration of a 2% solution of potassium chloride (KCl)?

2% solution = 2 g of KCl dissolved in 100 ml, or 20 g dissolved in 1 l

1M solution of KCl = 74.5 g dissolved in 1 l

(molecular weight of KCl = 74.5 g; RAM of K = 39, Cl = 35.5)

1 g dissolved in 1 l is = $\dfrac{1}{74}$ M

Therefore: 20 g dissolved in 1 l is = $\left(\frac{1}{74}\right) \times 20$ M = 0.27 M KCl

2.5.1 Calculating the molecular weight of a substance

Step 1: First find the formula of the substance (if not already given).
Step 2: Look up the atomic weights of the elements of the substance (if not already given).
Step 3: Add up all the individual weights of the elements and molecules (multiply the atomic weights by the number of atoms; if there is a number outside the bracket multiply everything inside the bracket by this number).
Step 4: Check your calculation.

Example 2.7

Given these atomic weights: Hydrogen (H) = 1.01; Copper (Cu) = 63.57; Carbon (C) = 12.01; Oxygen (O) = 15.99; Sulphur (S) = 32.07:

a) Find the molecular weight of copper sulphate ($CuSO_4$).

Molecular weight = 63.57 + 32.07 + (4 × 15.99) = 159.6 g

b) Find the molecular weight of lactic acid ($CH_3CHOHCOOH$)

Molecular weight = 6H + 3C + 3O = (6 × 1.01)

$\qquad$ + (3 × 12.01) + (3 × 15.99) = 90.06 g.

2.6 Calculations involving solutions

2.6.1 Calculating the amount of solute needed for making a given stock solution

To make a 25 ml stock solution of 0.1 M copper sulphate:

Step 1: First find the molecular weight of copper sulphate, which will give the weight of 1 mol of the substance, i.e. 159.6 g.

Step 2: A one molar solution of copper sulphate would be 1 mol dissolved in 1000 ml of water, i.e. 159.6 g dissolved in 1000 ml.

Step 3: A litre of 0.1 M solution would be made by dissolving one-tenth of the molecular weight (i.e. 15.96 g) in 1000 ml of water.

Step 4: But since only 25 ml of water is going to be used, only a fraction of the weight ($\frac{25}{1000}$) of copper sulphate is needed, i.e.

$$15.96 \times \frac{25}{1000} = 0.399 \, g.$$

Alternatively, using the formula: we know that the molar concentration = mol/volume, where mol = mass of solute/molecular weight. Or: concentration = (mass of solute/molecular weight)/volume.

Thus,

$0.1 \, M = $ (mass of solute/159.6 g)/0.025

or $0.1 \times 0.025 = $ mass of solute/159.6

Mass of solute required $= 0.1 \times 0.025 \times 159.6 = 0.399 \, g$

2.6.2 Concentration as proportions

Concentration can also be expressed as proportions, for example, parts per thousand (*ppt*) or million (*ppm*). If 1 mg of NaCl is dissolved in 1 l of distilled water (DW), the concentration of NaCl can be expressed as a 1 ppm solution.

1 ppm = 1 part of NaCl in 1 million parts of water

Assuming a specific gravity of 1 for water, 1 g of water has a volume of 1 ml.

1 ml = 1 g; 1 l = 1000 g

1 g of NaCl in 1000 g DW → 1 ppt

31

$$1 \text{ mg of NaCl in } 1000 \text{ g DW} \rightarrow 1 \text{ ppm}$$

$$\text{Hence, } 1 \text{ ppm} = 1 \text{ mg l}^{-1}$$

Measuring parts per million of solutes is important in techniques such as gas chromatography (Chapter 4). mg l^{-1} is equivalent to µg ml^{-1} because the units mg and l are divided by 1000 to give µg and ml respectively.

Equivalents

$$1 \text{ ppm} = 1 \text{ mg l}^{-1}$$

$$1 \text{ ppm} = 1 \text{ µg ml}^{-1}$$

$$1 \text{ ppm} = 1 \text{mg dm}^{-3}$$

$$(1 \text{ dm}^3 = 1 \text{ l})$$

2.6.3 Percent solutions

Many chemical solutions are mixed as percent concentrations. For example, 0.9% NaCl solution is 0.9 g NaCl in 100 ml water.

Weight/volume

Weight/volume or w/v means weight in grams of solute per volume of solution. When working with a dry chemical, the percent is calculated as grams of dry chemical in 100 ml. This is known as a weight/volume or w/v formula. For example:

- A 12% solution weight to volume (w/v) = 12 g in 100 ml or 120 g in 1 L.
- If 2 g is dissolved in 25 ml of water, the percentage solution is $(\frac{2}{25}) \times 100 = 8\%$
- Blood glucose concentration = 80 mg per 100 ml, i.e. 80% w/v; if expressed in mg per l, it would be 800 mg l^{-1}.

Volume by volume or % v/v

When the formula includes a liquid chemical, then the percent solution is the volume added of the chemical in ml made up to 100 ml with solvent. This is also a volume by volume (v/v) dilution. It means volume in ml of solute per volume of solution.

For solutions with two liquids, then % (v/v) is used.

Example 2.8

a) 10% acetic acid solution in water is 10 ml glacial acetic acid added to 90 ml water.
b) A solution containing 15 ml of solute in 100 ml of water would have a concentration of 15% (v/v).

Weight by weight or % w/w

Solutions can be either w/w (weight by weight) or w/v (weight by volume).

Since water is normally used, and 100 g of water is virtually the same as 100 ml, percent solutions using water can be given as either w/w or w/v.

Example 2.9

10% w/w (or w/v) solution of sodium chloride (NaCl) in water would contain 10 g NaCl dissolved in 100 g *or* 100 ml of water.

2.6.4 Final concentrations

We have previously considered how to make stock solutions by weighing out a known mass of solute. Sometimes it may be necessary, particularly in pharmacology or biochemistry practical work, to make up a more concentrated solution of stock than the final concentration in the sample (which will be less concentrated due to dilution).

Example 2.10

You are preparing a series of concentrations to test a drug on animal cells (e.g. rat liver cells) suspended in a buffer solution of a 5 ml total volume.

Prepare an initial stock solution of high concentration (e.g. 1 M) which can be used in a *serial dilution* (see page 26) to prepare the following dilute stock solutions: 1 M, 0.1 M, 0.01 M, 0.001 M and 0.0001 M (this can also be written as: 1 M, 1×10^{-1} M, 1×10^{-2} M, 1×10^{-3} M, 1×10^{-4} M).

Next, using $C_1 V_1 = C_2 V_2$, different volumes of the dilute stock solutions are used to dilute the buffer up to the concentrations desired.

Calculate the volume of dilute stock solution needed to add to the buffer (total volume of 5 ml).

e.g. if the final concentration of the drug in the buffer has to be $1\,\mu M$ ($= 1 \times 10^{-6}$ M), you can use the dilute stock solution of concentration 1×10^{-3} M to dilute the buffer.

using $C_1 V_1 = C_2 V_2$:

$$(1 \times 10^{-3}) \times V_1 = (1 \times 10^{-6}) \times 5$$
$$V_1 = (5 \times 10^{-6})/10^{-3};$$
$$V_1 = \left[5 \times 10^{-6 - (-3)}\right]$$
$$= \left[5 \times 10^{(-6+3)}\right] = 5 \times 10^{-3}$$
$$V_1 = 5\,\mu l$$

3

Separation of Liquids and Solids

Learning outcomes

- To perform filtration.
- To perform paper and column chromatography.
- To be able to use a centrifuge.
- To separate substances by gel electrophoresis.

3.1 Filtration

Filtration is a method of separation of liquids and solids from a mixture of liquid or gas by passing it through a filter. The liquid (or gas) which passes through is called the *filtrate*, while the solid remaining on the filter is called *residue*. This technique is useful in removing impurities or for isolating solids. Two common types of filtration used for these purposes are *gravity* filtration and *vacuum* or *suction* filtration.

The types of filter that are used can vary, but commonly we use filter paper and glass fibre filters of various pore sizes. Glass fibre filters allow fast flow rates and they have high-loading capacity, wide thermal tolerance and excellent precipitate retention. They are used to filter proteins, cells, air particles and water particles.

3.1.1 Gravity filtration

The usual method employed in gravity filtration is as follows:

Step 1: Use the right size filter paper for the funnel that you will be using (Figure 3.1a), so that the folded paper will rest a few millimetres below the rim of the funnel.

Step 2: Fold your filter paper in half to produce a half moon shape. Then fold once more to create a triangle. Open the paper,

---

Essential Laboratory Skills for Biosciences, First Edition. M.S. Meah and E. Kebede-Westhead.
© 2012 John Wiley & Sons, Ltd. Published 2012 by John Wiley & Sons, Ltd.

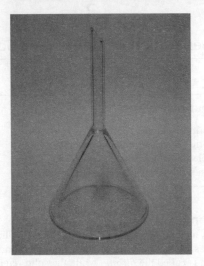

Figure 3.1a Funnel

Figure 3.1b Cone-shaped filter paper

which will now resemble a cone (Figure 3.1b). If a fluted paper is required to increase surface area, from the half moon shape, fold each half into quarters, and then into eighths until all folds possible are done (Figure 3.1c).

Figure 3.1c *Fluted-shaped filter paper*

Step 3: Place a glass funnel in a ring or in an Erlenmeyer flask (see Figure 1.2b in Chapter 1). Wet the filter paper with a few millilitres of the solvent to be used in order to keep the paper tightly against the glass funnel.

Step 4: Pour the mixture into the neck of the funnel, a little at a time, until all the liquid is in the second flask. Discard the used filter paper in an appropriate waste container if the residue is not of interest.

3.1.2 Vacuum filtration

Vacuum filtration is useful to collect a particular solid. The mixture of solid and liquid is poured through a filter paper in a Buchner funnel (Figure 3.2). The solid is trapped by the filter and the liquid is drawn through the funnel into the flask below by a vacuum.

Vacuum filtration is faster than gravity filtration because the fluid and air are forced through the filter paper by using reduced pressure.

The equipment required for vacuum filtration is as follows:

- Buchner funnel (Figure 3.2); also sintered, glass and metal filter holders and funnels available for microanalysis
- Heavy-walled, side arm filtering flask (Figure 3.3)
- Rubber adaptor or stopper to seal the funnel to the flask when under vacuum
- Vacuum source

Figure 3.2 Buchner funnel

Figure 3.3 Side-arm filtering flask

⚠ Reduced pressure will mean that liquids with a low boiling point (<125°C) may boil off and be lost.

The procedure for vacuum filtration is as follows:

Step 1: Attach parts of the apparatus. Clamp the flask securely to a ring stand. Add a rubber adaptor, then a Buchner funnel on the adapter, then a filter paper in the funnel – checking that it will stay flat.

Step 2: Using thick tubing, connect the side arm of the flask to a vacuum source (this can be mechanical, or alternatively a vacuum can be created with a water trap plus a water aspirator).

Step 3: Place a small amount of the liquid on the filter paper (this causes the paper to attach more tightly to the base of the funnel) and then turn on the vacuum. Check that air is being pulled through the filter paper – briefly placing a hand on top of the Buchner funnel will tell you if there is suction.

Step 4: Pour the mixture to be filtered onto the filter paper (ideally in the centre). The vacuum will pull the liquid through the funnel. Check that particles are not escaping under the filter paper.

Step 5: Rinse the flask with more liquid to remove any remaining mixture, and pour this onto the filter paper. Make note of volume of liquid used if it is quantitative filtration.

Step 6: Remove the rubber tubing before turning off the vacuum source.

Step 7: Carefully remove the filter paper and the filtered solid that is on it.

Step 8: Place it on a watch glass and let it dry in the air.

Step 9: If the filter is clogged, you may need to start a new filtration and, if possible, to pre-filter the sample through a filter with a larger pore size.

Note:

If the filter is clogged with too much solids retained on it, you may be tempted to compensate by increasing the vacuum pressure, but this is inadvisable because it may cause the filter to break.

3.2 Centrifugation

Centrifugation is a method of separating liquids or solids from a fluid. It uses an instrument called a centrifuge (Figure 3.4a) to spin the fluid

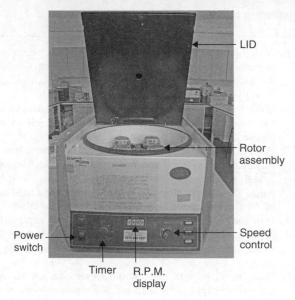

Figure 3.4a Low-speed centrifuge

at high forces and at a fast speed in order to separate the substances mechanically.

The centrifuge works on the principle of using a motor to accelerate the fluid in a tube, that is by rotating or spinning it. The centrifuge produces what is commonly referred to as 'centrifugal force', which it uses to separate the substances. What we call centrifugal force is actually the result of what happens when a particle is accelerated away from a central point but is unable to move beyond the perimeter and has to move instead in a circular motion. It behaves as though there is a force causing the particle to move away (i.e. outward) from the axis of rotation. The heavier the particle, the stronger its compulsion to try to move outward.

The acceleration due to gravity is called g (earth's gravitational force) and has the value $9.8\,\mathrm{m\,s^{-2}}$. A centrifuge can produce multiples of g (called the relative centrifuge field or RCF), depending on the speed of revolution in revolutions per minute (rpm), the length and the radius of the rotor. The bigger the rotor length or the speed of rotation, the greater the virtual g force. This can be expressed in the relationship shown below:

$$g(\text{or RCF}) = 1.118 \times r \times 10^{-5} \times (\text{rpm})^2$$

where r = radius of the arm of the rotor in cm; (rpm) = revolutions per minute of the rotor.

Other factors to take into account in separation of the substances include the size and shape of particles, viscosity of liquid, density of particle, and concentration of solution.

3.2.1 Types of centrifugation

Differential (sedimentation)

Mixtures of different-sized particles are separated so that the solid particles are concentrated from the liquid. The heavier particles will move towards the bottom of the tube due to the apparent centrifugal force. By adjusting the g value, different substances can be separated.

Examples of this include separating organelles from cells in solution or separating cells from blood.

Density gradient

A centrifuge tube has a solution which increases in density from the top to the bottom. This will allow separation by two mechanisms:

(i) Larger particles will move faster through the gradient than the smaller ones.
(ii) Particles with higher density will be at the bottom of the tube and those with lower density will be at the top of the tube.

3.2.2 Types of centrifuge

A typical centrifuge will consist of a motor which will spin a rotor arm (usually there are four or six – see Figure 3.4a) to which attaches a centrifuge tube. The rotors are usually one of three types – swing-out rotors, fixed angle or vertical – depending on the type of centrifuge. The centrifuge tubes are made of glass or plastic, and vary in size depending on the volumes of solution to be used (from $2.5\,\mu l$ to $1000\,ml$).

There are different sizes of centrifuges, depending on what substances are to be separated. These range from low-speed centrifuges (Figure 3.4a) with speeds up to 6000 rpm for separating cells and larger organelles like nuclei, to high-speed (up to 25 000 rpm) centrifuges which are used to separate cells and smaller organelles (Figure 3.4b). Finally, there are

Figure 3.4b High-speed centrifuge

high-speed ultracentrifuges producing up to 30 000 rpm and 600 000 × g, which are used to separate very small organelles such as ribosomes.

⚠ Balancing the rotor arms of centrifuges is essential, as any imbalance at high speed and force may damage the centrifuge, and in worst cases, could cause an accident.
Always make sure the lid is closed properly.

Applications of centrifugation include the separation of blood into plasma and red blood cells, sedimentation of cells and viruses, separation of sub-cellular organelles, and isolation of macromolecules such as DNA, RNA, proteins or lipids.

3.2.3 Example procedure
The following describes an example procedure to separate a blood sample using a microhaematocrit centrifuge (Figure 3.5).

SEPARATION OF LIQUIDS AND SOLIDS

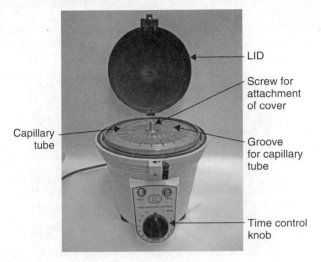

Figure 3.5 *Microhaematocrit centrifuge*

Separating a blood sample using a microhaematocrit centrifuge

Step 1: Select a clean glass capillary tube or centrifuge tube (Figure 3.6a).

Step 2: Place one end of the capillary tube a few mm into a tube containing blood at an angle of approximately 45°. Allow blood to rise up by capillary attraction to at least 50–70% of the capillary tube.

Step 3: Place the blood end of the capillary tube into the sealing medium (Cristaseal) and ensure that a small sealed plug is formed.

Step 4: Place the capillary tube into the circular grooved tray of the centrifuge, ensuring the sealed end is outward. Each numbered groove takes one capillary tube (Figure 3.6a).

Step 5: Ensure that another capillary tube is placed opposite the first one.

Step 6: Close the lid and set the time (e.g. 3 minutes) and the g force (e.g. $3000 \times g$).

Step 7: Examine the capillary tube (Figure 3.6b). You should see straw-coloured plasma at the top and red blood cells at the bottom (towards the sealed end).

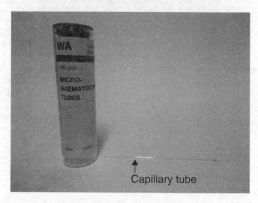

Capillary tube

Figure 3.6a Capillary tube

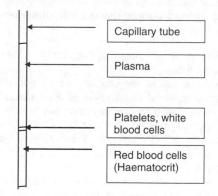

| Capillary tube |
| Plasma |
| Platelets, white blood cells |
| Red blood cells (Haematocrit) |

Figure 3.6b Capillary tube showing separated components after centrifugation

 Make sure you wear gloves when handling blood.

3.3 Chromatography

The two common methods of separating compounds from mixtures are *chromatography* and gel *electrophoresis*.

The principle behind chromatography is that solutes in a mixture have different *affinities* (attractions) for particles of an insoluble matrix over which a solution of the components is passing. The insoluble matrix is called the *stationary phase* (solid or liquid on a solid), while the solution which passes through it is called the *mobile phase* (liquid or gas).

The mobile phase flows through the stationary phase and carries the components of the mixture with it. Different components travel at different rates. Separation depends on *size, charge*, and the type of *substance* (e.g. proteins, peptides) which needs to be separated, and also the type of *matrix* used and therefore the type of chromatography. The matrix can be:

- paper – *paper chromatography*
- solid matrix – placed in a column called *column chromatography*
- glass, aluminium or polymer sheet – *thin layer chromatography*.

Once chromatography has been used to separate a mixture, then a variety of methods are available to detect the separated components, depending on the type of chromatography.

3.3.1 Paper chromatography

In paper chromatography, the stationary phase is a very uniform absorbent paper. The mobile phase is a liquid solvent or a mixture of solvents. Separation of components depends on their solubility in the mobile phase, and their different attractions in the mobile and stationary phase.

The principles behind paper chromatography are:

- *Capillary Action* – the movement of liquid within the spaces of a porous material due to the forces of adhesion, cohesion, and surface tension. The liquid is able to move up the filter paper because its attraction to itself is stronger than the force of gravity.
- *Solubility* – the degree to which a material (solute) dissolves into a solvent. Solutes dissolve into solvents that have similar properties. This allows different solutes to be separated by different combinations of solvents.

Amino acids are commonly separated using paper chromatography. The following example shows how this can be done to separate the amino acids alanine and glutamine:

Step 1: Wearing gloves, obtain a sheet of chromatography paper approx. 15–25 cm wide. Note: avoid touching the paper with fingers, since oils from your skin will affect the movement of molecules on the paper.

Step 2: Draw a pencil line 3 cm from the bottom of the paper.

Step 3: Draw tick marks at 2.5–3 cm intervals from the left end of the line and label.

Step 4: Pipette a 5 µl volume onto the paper to 'spot' the mixture to be separated on the pencil line (touch the paper with a small drop of size 2–3 mm diameter). Let it dry and then repeat, adding drops to the same spot.

Step 5: Repeat Step 4 for the other remaining solutions and label these, e.g. 'Known amino acids', 'Unknown amino acids'.

Step 6: Take a container and place a small amount of solvent, consisting of butanol, acetic acid and water in the ratio 2:1:1.

Step 7: Fold the chromatography paper into a cylinder held by paper clips, and place in the container so that the solvent is below the pencil line and the tick marks.

Step 8: Cover the container to ensure that the atmosphere inside it is saturated with solvent vapour, and keep it out of direct sunlight and heat for approx. 1.5–2 hours. As the solvent slowly travels up the paper, the different components of the mixtures travel at different rates and the mixtures are separated.

Step 9: Remove the paper and mark the solvent front, then dry it in an oven (120°C) or a heater (in a fume cupboard, solvents are commonly volatile).

Step 10: Spray the paper with Ninhydrin solution; this causes the amino acids to show up as coloured spots. Then dry it with warm air. This final result is called a 'chromatogram' (Figure 3.7).

Step 11: Circle the spots with a pencil and measure the distance each spot has travelled from the origin (pencil line) and the distance travelled by the solvent.

Step 12: Calculate the R_f (Retardation or Retention factor):

$$R_f = \frac{\text{distance moved by substance}}{\text{distance from origin to solvent front}}$$

Step 13: Compare the R_f values and determine the unknown substances with the known substances.

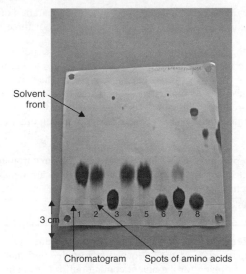

Figure 3.7 *Chromatogram showing baseline, spots and solvent front*

3.3.2 Thin layer chromatography

In thin layer chromatography, the stationary phase is a powdered adsorbent (silica gel or alumina) which is fixed to aluminium, glass, or plastic plate. The mixture to be analyzed is loaded near the bottom of the plate. The plate is placed in a reservoir of solvent so that only the bottom of the plate is submerged. This solvent is the mobile phase; it moves up the plate, causing the components of the mixture to separate into 'spots'.

3.3.3 Column chromatography

In column chromatography, the stationary phase is a powdered adsorbent (silica gel or alumina) or beads which are placed in a vertical glass column (variable in size). The mobile phase is a solvent poured on top of the loaded column. The solvent flows down the column, causing the components of the mixture to separate and elute out at different rates, depending on size or shape, charge or affinity. For example, large molecules travel faster than small molecules (since smaller molecules are slowed by the matrix) and therefore come out of the column first. Spherical molecules move through the matrix more slowly than narrow, rod-shaped molecules and therefore they elute later.

The following is a simplified procedure for the separation of compounds by column chromatography. The apparatus consists of three main parts:

- Chromatography **column** – a tube containing a porous disc which keeps the matrix in and allows water and solutes to get through.
- **Matrix** – consists of beads having tiny pores. The size of pores determines the range of molecular weights that the matrix can separate.
- **Buffer** – carries the sample through the matrix. It is a solution of constant pH which resists changes in the pH when other chemicals are added.

The main steps in processing samples through column chromatography are highlighted below. However, the detailed steps are dependent on the type of chromatography and equipment used for analysis, and this is beyond the scope of this book.

Step 1: Remove any existing buffer in the column. Check that the appropriate matrix is loaded or installed and that a collector is in place for the eluent coming out of the column.

Step 2: Load an appropriate amount of sample (e.g. 0.2 ml) to the column, and open the valve or start method (in the case of automated equipment) to allow the sample to enter the matrix.

Step 3: Add buffer to cover the matrix.

Step 4: Place a tube under the column and collect the eluent.

Step 5: Repeat the collection of fluid with further tubes. The collected fluid samples may need to be assayed further to distinguish the type of substance.

There are various forms of column chromatography, where substances are separated by size exclusion, ionic charge and affinity chromatography. Basic differences between these different chromatography methods are explained below.

Size exclusion chromatography (SEC)

Also called gel filtration chromatography, SEC involves separation of substances through a porous column (gel such as agar and polyacrylamide or silica), based on the size of molecules. Molecules smaller than the pores penetrate the matrix and thus take a longer time to travel through the column compared with larger molecules that pass through the column but do not penetrate the pores. The larger molecules are thus 'excluded' according to size.

This method is commonly used in the separation of biomolecules such as proteins, polysaccharides and nucleic acids. The process is passive and the molecules do not interact with the column matrix.

Ion exchange chromatography (IEC)

IEC involves the separation of solutes based on their electrical charges, where charged solutes interact (adsorption) with oppositely charged molecules in an ion exchange column and become bound or immobilized. A positively charged column (anion exchanger) binds negatively charged compounds, while a negatively charged column (cation exchanger) binds positively charged compounds.

Two main steps are involved:

(i) binding of solutes (adsorption); and
(ii) elution of the various compounds at different stages to separate the components.

The second step, where solutes are removed from the column (desorption), is achieved by changing the ionic strength of the mobile phase or the pH of the elute buffer. The solutes with the weakest ionic concentration and strength of binding are desorbed and eluted first, while solutes with stronger ionic concentrations elute later because they have a greater interaction with the column matrix and therefore require higher salt concentration for displacement. Detection of the solute components can be made using UV or visible-light spectrophotometry or conductivity.

IEC is commonly used to separate and purify charged biomolecules (proteins, polypeptides, nucleic acids, etc.) in water analysis (nitrate, phosphate, sulphate, etc.) and quality control.

Affinity chromatography

In affinity chromatography, similar to ion exchange chromatography, solutes are separated based on their affinity to compounds in the column matrix. In this case, it involves specific interactions such as between antigen and antibody, or enzyme and substrate, or biomolecules (e.g., amino acids) and metals (e.g., copper, zinc), often involving strong covalent bonding. Compounds with low affinity are bound loosely to the matrix and can be washed off the column or eluted, while those with higher affinity get bound to the column matrix. Compounds of interest are eluted by changing the pH, by increasing the strength of the buffer (mobile phase), or by adding substances that can compete for affinity or binding sites in the column matrix.

The method is commonly used to purify a compound from a mixed sample, e.g. a protein mixture.

3.3.4 Gas chromatography (GC)

Separation of mixtures in gas chromatography involves the same principles as in column chromatography, but a major difference is that it requires the sample to be vaporized. It is not suitable for labile compounds that decompose on heating, or reactive chemicals like metabolites. It is used to separate mixtures such as hydrocarbons (e.g. crude oil, gasoline), short chain fatty acids or alcohols (e.g. in tests for drink driving).

In contrast to column chromatography which has a liquid carrier and a solid stationary phase, the carrier or the mobile phase in GC is an inert gas such as helium or nitrogen, and the stationary phase is a high-boiling liquid – generally a waxy polymer of silicon which is thinly packed into a long, narrow glass or metal column. The mixture to be analyzed is loaded by syringe into the column, where it is first vaporized, and then carried through the column in the mobile phase (e.g. helium).

Achieving separation of the gaseous mixture depends on column temperature, boiling point and solubility of compounds in the liquid phase. Separated compounds flow through a flame ionization detector or a conductivity detector at the end of the column, and successful separation shows specific peaks for each compound.

GC is also used in conjunction with mass spectrophotometry (GC-MS) or infra-red spectroscopy (GC-IR) for detection of the gaseous products.

Detection of separated compounds in column chromatography

In the chromatography methods described above, compounds separated from a mixture are either collected and analyzed further to identify and/or quantify, or are detected as they are eluted from the column. There are several modern automated equipment where a sample is injected at one end of a column and the separated compounds are commonly detected with conductivity (ion exchange chromatography) or UV or visible-light absorption conductivity (most column chromatography). Most organic compounds absorb UV light of various wavelengths, and different compounds can be identified over the range of the UV spectrum.

3.3.5 High performance liquid chromatography (HPLC)

HPLC is basically a refined automated column chromatography. It is made more sensitive in that instead of gravity force (as in the methods described above), high pressure is used to move both mobile phase and sample through a column. The column is densely packed with smaller particles, which increase the surface area for interaction of molecules with the stationary phase. This allows finer separation of mixtures within a short distance. Separation of mixtures is based on affinity to the mobile phase (commonly methanol, acetonitrile).

Detection, usually by UV/visible spectrophotometry, is highly automated and sensitive, enabling quantification of separated compounds from the peaks generated. Detection is also made by channelling eluents through a mass spectrometer coupled to the HPLC.

3.4 Electrophoresis

Some molecules carry a charge, the size of which depends on the type of molecule, the pH and the medium the molecule is in. If an electric field is applied, charged molecules in solution move to the electrodes of opposite charge (i.e. negatively charge molecules move towards the positive electrode or *anode*, while positively charged molecules move towards the negative electrode or *cathode*). This separation of molecules based on the type of charges they carry is called *electrophoresis*. Other factors which are important in this movement are the size and shape of the molecules.

This technique is useful in separating proteins in blood (e.g. albumin, transferrin). It is also useful in identifying abnormal antibodies. Another use is to identify isoenzymes (such as those of lactic dehydrogenase), which can then be used in diagnosing any damage to the heart, such as from a heart attack. The separated substances are stained to identify them and can be measured by assay.

There are various types of electrophoresis, which depend on the medium the molecules are in. These include paper, cellulose acetate, starch-block, polyacrylamide-gel electrophoresis (PAGE), sodium dodecyl sulphate (SDS), gel electrophoresis and gradient-gel electrophoresis. The methods which are currently most popular are those using gels. A jelly-like substance allows small molecules to get through easily, but large ones have difficulty.

SDS is a detergent which binds to proteins easily. In combination with PAGE, it has become a popular method of separation of proteins because the detergent masks the surface charge, and thus the proteins move according to size, with the larger ones moving slowly and the smaller ones faster. An example of a gel method of electrophoresis is shown below.

3.4.1 Gel electrophoresis

This technique separates molecules based on their charge, size and shape. The principle is as follows: a mixture is applied to a gel and an electric current is placed across the gel. The current causes the charged molecules

51

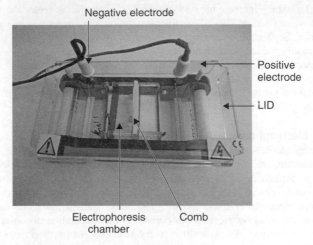

Figure 3.8 *Gel electrophoresis chamber*

of the substance to move either to the positive or negative poles. The direction of movement, the speed and the distance travelled are related to the charge, shape and size of the molecules.

The apparatus required for gel electrophoresis comprises:

- **An electrophoresis chamber**: a rectangular clear plastic hollow chamber containing buffer with electrodes. The longest sides have 4–5 equally spaced notches, to which are attached plastic combs to separate the chamber (Figure 3.8).
- **Gel**: made by dissolving agarose powder in hot buffer; when it cools, it solidifies into a gel which has numerous pores.
- **Buffer**: this conducts electricity and controls the pH. Without it, the stability and charge of samples would alter.
- **Samples**: mixtures of chemicals. Small molecules move more easily through the gel than large molecules.
- **Power supply**: provides current in the gel, which causes negative charges to move towards the positive electrode and positive charges towards the negative electrode. Speed of movement is proportional to voltage.

The following procedure describes the separation of organic molecules by gel electrophoresis:

Step 1: Cover the ends of the electrophoresis chamber with rubber end-caps.

Step 2: Attach the plastic comb in the middle notches so that there is a space between the bottom of the teeth and the base of the chamber.

Step 3: Mix a 0.8% w/v mixture of agarose powder in buffer.

Step 4: Heat the mixture until the agarose dissolves, then let it cool. When the solution is cooled to 50°C, pour it into the chamber.

Step 5: When the gel solidifies, remove the comb, remove the rubber end caps and this will leave small rectangular wells formed by the comb in the gel (Figure 3.9). Submerge this gel under the buffer.

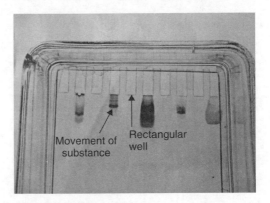

Figure 3.9 Gel electrophoresis wells

Step 6: Using a pipette, load samples into the wells, including known standards as well as unknown samples.

Step 7: Place the cover on the electrophoresis chamber, making sure the red and black plugs are attached correctly. Then connect the power supply (red wire to red socket and black wire to black socket).

Step 8: Switch on the power supply and set voltage to 90 V.

Step 9: After 30 minutes switch off and note the distance moved by the substances (Figure 3.9).

4

Common Techniques and Equipment

Learning outcomes

- To know the principles of common techniques.
- To use a pH meter.
- To perform titration using an indicator or pH meter.
- To use a spectrophotometer.
- To know the common aseptic techniques.
- To use common aseptic techniques (flame a loop, streak a plate).

4.1 Titration

4.1.1 Acids and bases

Acids are substances which release protons (H^+ ions). Bases are substances which release hydroxyl (OH^- ions) or absorb H^+ ions (Table 4.1). Adding extra protons increases the acidity of a solution, while adding extra hydroxyl ions increases the alkalinity. The strength of an acid or base is determined by how readily they dissociate to give up protons or release hydroxyl ions. Strong acids are almost completely dissociated, while weak acids are partially dissociated. Similarly, strong bases are almost completely dissociated and weak bases are partially dissociated. A measure of the dissociation can be represented by the equilibrium constant for an acid or a base.

The acid dissociation constant (Ka) is a measure of the acid dissociation; a high value would indicate a strong acid. However, since Ka values are very small, we convert them to logarithmic values (pKa), with which we can work more conveniently like the pH scale (i.e pKa $= -\log$ Ka). The lower the pKa is, the stronger the acid.

Essential Laboratory Skills for Biosciences, First Edition. M.S. Meah and E. Kebede-Westhead.
© 2012 John Wiley & Sons, Ltd. Published 2012 by John Wiley & Sons, Ltd.

Acids react with bases to produce salt and water, i.e.

$$Acid + Base \rightarrow Salt + Water$$

e.g. $HCl + NaOH \rightarrow NaCl + H_2O$

Hydrochloric Acid + Sodium Hydroxide $\rightarrow$ Sodium Chloride

+ Water

By convention, the acidity and alkalinity of solutions is defined by a positive pH scale (1–14). This is a log scale without units, used to quantify the number of protons (H^+) in solution since we are dealing with very small values. pH stands for the potential of hydrogen ions and is the negative log of the concentration of H^+.

$$pH = -\log [H^+]$$

- The higher the H^+ concentration, the lower the pH. Hence: pH = 1 (strong acid), pH = 14 (strong base), pH = 7 (neutral solution, like water).
- Pure water has H^+ concentration of 10^{-7} M giving a pH of 7. Arterial blood has a pH of 7.4.
- Stomach acid (HCl) has H^+ concentration of 10^{-1} M, giving a pH of 1.
- A 10 fold change in H^+ concentration will cause the pH to change by 1.

$$pH = 1$$
$$pH = -\log 10^{-1}$$
$$pH = -(-1) \log 10$$
$$pH = +1 \times \log 10$$
$$pH = 1$$

Example 4.1

If the H^+ concentration is 3.5×10^{-3} M, what is the pH?

$$pH = -\log [H^+]$$
or $pH = -\log (3.5 \times 10^{-3})$

log calculation:

$$pH = -(-2.46)$$
or $pH = 2.46$

Table 4.1 *Acids and bases commonly used in the laboratory*

Strong acid	Weak acid
Hydrochloric acid (HCl)	Carbonic acid (H_2CO_3)
Nitric acid (HNO_3)	Ethanoic acid (CH_3COOH)
Sulphuric acid (H_2SO_4)	Methanoic acid (HCOOH)
Strong base	**Weak base**
Sodium hydroxide (NaOH)	Ammonium hydroxide (NH_4OH)
Potassium hydroxide (KOH)	Ammonia (NH_3)
Lithium hydroxide (LiOH)	Methylamine (CH_3NH_2)

Example 4.2

If the pH is 7.4, what is the H^+ concentration?

$$pH = -\log [H^+]$$
$$or\ 7.4 = -\log [H^+]$$
$$or\ \log [H^+] = -7.4$$
Taking antilog,
$$[H^+] = 3.9 \times 10^{-8}\,M$$

4.1.2 Measuring pH

Electric pH meters or pH papers are used to measure pH. There are also battery-operated pH meters for field use.

pH paper

pH paper is coated with a chemical indicator which changes colour depending on the concentration of H^+ ions. There are different pH papers, measuring the whole range (1–14) or various ranges of pH (see Figure 4.1).

Examples of pH paper types include:

- Litmus paper will change from red to blue if it is in contact with an acidic solution (pH = 4.5–8.3)
- Phenol red will change from yellow to red in the pH range 6.8 to 8.0.

Figure 4.1 pH paper

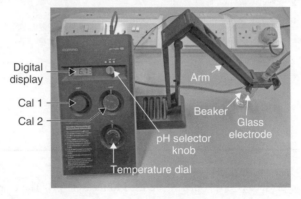

Figure 4.2 pH meter

Electric pH meters

There are digital or analogue types of electric pH meters (see Figure 4.2).

The electric pH meter consists of an electric meter connected to a glass electrode filled with potassium chloride (KCl). The glass surface is very thin and permeable to protons. There is a platinum wire inside the glass electrode. Current is carried from the glass surface through the KCl solution to the platinum wire, and the voltage produced is read by the electric meter. This voltage is proportional to the pH of the solution into which the electrode is placed. Changes in temperature can affect the resistance of the glass bulb; this can affect the pH reading, so it has to be taken into account in the pH measurement.

Using a pH meter
It is always essential to calibrate pH meters before use. This is usually done using two buffers – one at pH 4 and the other at pH 7.

Calibration of pH meter

Step 1: Two beakers are prepared with 50 ml of distilled water each, to which are added buffer tablets so that the pH in each is 4 and 7 respectively. It is important that solutions are mixed thoroughly for all measurements. This is typically done using a magnetic stirrer.

Step 2: Note the temperature of the buffer and input this into the pH meter using the temperature dial.

Step 3: Place the glass electrode into the buffer at pH 7.

Step 4: Adjust the digital 'Zero' knob or input button 'Cal 1" (see Figure 4.2) so that the value reads 7.00.

Step 5: Lift the electrode and wash with distilled water from a wash bottle.

Step 6: Place the electrode in buffer at pH 4 and adjust the other digital knob ('Cal 2') to set to 4.0.

Step 7: Lift out the electrode and wash as before, then put into the buffer at pH 7 and check that the reading is 7.00. You can also check the buffer at pH 4, following the method described for pH 7.

Step 8: Wash the electrode with distilled water and now place in the sample liquid. If the sample temperature varies, adjust the temperature before taking a reading.

4.1.3 Buffers
Buffers resist changes in pH when acids or bases are added.

Many enzymes have a pH range above or below which they are not active. In a living organism, cells keep the pH constant by using buffers which are weak acids or bases (e.g. bicarbonates or phosphoric acid). Typically, a buffer is a weak acid and the salt of a weak acid (e.g. carbonic acid + sodium bicarbonate) or a weak base and the salt of that base (e.g. ammonia + ammonium chloride). These substances can 'mop up' excess protons or bases by producing opposite charges to those added.

To see how buffers react to alterations in acids and bases, a titration curve can be performed. Using a burette (see Figure 1.2d in Chapter 1), small amounts of a strong base are added to a buffer (a weak acid). There will be a point at which there is very little change in pH (buffering region), and the value of this pH is a constant called the pKa.

The relationship between pH, pKa, base and acid for a buffer system can be summarized by the *Henderson-Hasselbalch equation*, i.e.

$$pH = pKa + \log [base]/[acid]$$

This is an important relationship which shows how altering the acid base concentration will alter the pH. An important example of its use is in keeping arterial blood at a pH of 7.4 by the bicarbonate buffer system. In the above equation, pKa is 6.1 and the base is sodium bicarbonate and the acid is carbonic acid. To keep the pH at 7.4, the ratio of base concentration to acid concentration has to be 20:1.

4.1.4   Titration

Titration is a method which is used to determine the precise endpoint of a chemical reaction. A burette is used to deliver a measured volume of a reactant (called the *titrant*) to a flask which contains a known volume of the second reactant. An indicator or a pH meter is used to detect the endpoint of the reaction.

Procedure for titration using an indicator

Step 1: Fill the burette with the titrating fluid. Check for air bubbles and leaks.

Step 2: Place the solution to be analyzed in a conical flask or a beaker.

Step 3: Add a few drops of an appropriate indicator (e.g. phenolph-thalein).

Step 3: Note the volume in the burette and then use the burette's stopcock to dispense the titrant slowly into the flask.

Step 4: Slow down the delivery as you near the end point, which is shown by the colour change.

Step 5: Note the volume when the end point is reached and calculate the total volume used for the reaction.

4.2   Spectrophotometry

Spectrophotometry is a simple technique using absorption of light by substances in solution at specific wavelengths to determine their concentration.

The principle behind spectrophotometry is that different atoms, molecules and ions absorb specific wavelengths of light, so the amount of light absorbed by a sample is measured. The greater the light absorbance,

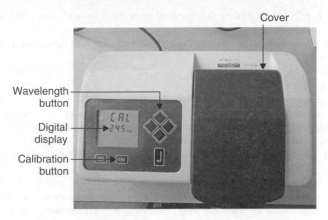

Figure 4.3 *Spectrophotometer*

the lower the amount of light transmitted. Absorbance is directly related to the concentration of the compound in the sample. This is represented by the *Beer-Lambert Law*, which relates the amount of light absorbance by a solution to the concentration of the compound in solution and the length of the light path (see Appendix 4, p. 127).

4.2.1 Apparatus

A spectrophotometer (Figure 4.3) is an instrument used to determine absorbance of light at different wavelengths; this is called *colorimetry*. Light radiation passes through a plastic, glass or quartz cell containing a 2–3 ml sample. The detector measures the energy after it has passed through the sample. The readout device calculates the amount of light absorbed by the sample and displays the signal from the detector as absorbance or transmission.

Components of a spectrophotometer include one or more light source(s), a wavelength selector, a sample holder and a device to measure light intensity (photodetector). In more detail, the equipment comprises:

- A **light source**: this can vary from ultraviolet to visible to infrared.
- A **filter**: typically a coloured glass plate, used to select specific wavelengths.
- A **photodetector**: this converts light energy into electrical energy, which is displayed on a meter. Most spectrophotometers will give absorbances up to 2.5. However, a linear relationship between absorbance and concentration is only strong up to an absorbance

of 1.0, above which the linearity may be lost and hence the line should not be extrapolated.

- Each **sample** to be analyzed is kept in a cuvette (see below) or a clear test tube.

- A **blank** contains the solvent (usually distilled water) used to dissolve the chemical you want to analyze, and this is used to calibrate the spectrophotometer. The blank has no colour and thus should not absorb any light. Where absorbance is measured after a colour reaction, a reagent blank is prepared containing the reagents and treated the same way as the samples (e.g., heating).

- **Cuvettes** (or **cells**): sample containers, which can be square or rectangular in shape. They are provided in pairs that have been carefully matched to allow transmission through the solvent and the sample. The most common transmission width is 1 cm, but narrower or wider cuvettes can be used for small volumes or for increasing sensitivity in solutions with weak signals, respectively. The type of cuvette used is determined by the wavelength to be measured. Commonly they are made of either optical or quartz glass or plastic. Some cuvettes have a small cap which is used to prevent loss of volatile solvents.

 Precautions in using cuvettes

- Cuvette pairs need to be regularly checked against one another to detect any differences from scratches and wear. Plastic cuvettes are routinely used in practicals, and they are not perfectly matched like the expensive quartz cuvettes. Hence, it is important not to switch cuvettes between blanks and samples in between measurements. Unmatched cuvettes can be a source of error especially if measuring at the lower end of the absorbance scale or detecting small differences between samples.

- Always clean the cells thoroughly and rinse at least once with a portion of the sample before filling with the sample for measurement.

- Always pick cuvettes up from the top by the side faces to avoid fingerprints.

- Always wipe the exposed surface of the cells dry and free from fingerprints, using tissue paper.

- Do not leave solutions in cuvettes for long, particularly in strong alkali.

- Ensure that there are no air bubbles on the inner surfaces of the cells.

- Never use a brush or any tool for cleaning, as this may scratch the optical surfaces.
- Make note of the sides of the cuvettes when placing in the sample chamber. Often they have opaque sides, and clear sides which should be aligned with the pathway of the light. Most cuvettes used in the lab routinely have an arrow from the top of the side which should be the side facing the light path (Figure 4.4).

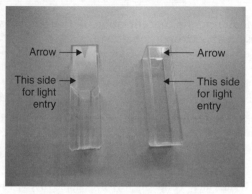

Figure 4.4 Cuvettes

4.2.2 Operating instructions for spectrophotometer

Using the Spectrophotometer

Step 1: Turn the spectrophotometer on and let it warm up for 15 minutes.

Step 2: Press the Transmittance/Absorbance selector switch.

Step 3: Adjust wavelength control to read the wavelength at which the absorbance is measured (e.g. 420 nm).

Step 4: Open the sample compartment and check the alignment of the light path before putting in the blank cuvette containing distilled water (Figure 4.3). Close the lid.

Step 5: Adjust absorbance reading to 0.000 by pressing 'Cal' or 'Set reference' on the panel.

Step 6: Remove the blank and put in cuvette with your sample.

Step 7: Read absorbance of sample to three decimal places.

Note: Remember to frequently check the calibration by using the blank and adjusting the absorbance back to zero. But do not press 'Cal' between samples.

4.2.3 Measuring the concentration of an unknown solution using a spectrophotometer

A series of standard solutions of known concentrations are used to determine the concentration of unknown solutions. The procedure is as follows:

Measuring the concentration of an unknown solution

Step 1: Prepare a stock solution of the compound to be determined. If you have a high concentration of the stock solution, you may need to lower the concentration, otherwise your absorbance values may be too high. For example, if you have 125 mM potassium dichromate, you can reduce this by doing a 1 in 10 dilution (1 ml of 125 mM mixed with 9 ml of distilled water) giving 12.5 mM, then a further 1 in 10 dilution to give 1.25 mM.

Step 2: Make a series of standard solutions of known concentrations using the stock solution (six different concentrations may be enough). It is advisable to prepare solutions with absorbance less than 1 (the relationship between absorbance and concentration may be non-linear above absorbance of 1).

Step 3: Measure the light absorbance of your standards after blanking with distilled water.

Step 4: Provided the solutions were diluted accurately, plotting absorbance against concentration should give a linear curve, as shown in Figure 4.5. This is called the standard or calibration curve.

Step 5: Measure the absorbance of the unknown samples and read the related concentration from the standard curve (Figure 4.5).

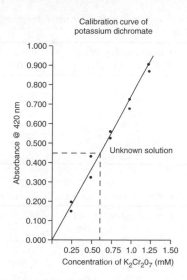

Figure 4.5 *Standard curve of light absorption against concentration*

For example, to measure the concentration of an unknown solution of potassium dichromate ($K_2Cr_2O_7$):

Measuring the concentration of an unknown solution of Potassium Dichromate

Step 1: Using a stock solution of 1.25 mM, prepare standards of known concentrations of 2 ml volume using distilled water for dilution (see Table 4.2).

Step 2: First decide on the concentrations of standard solutions (e.g. 0, 0.25, 0.50, 0.75, 1.0 and 1.25 mM).

Step 3: Calculate the volume of stock solution and distilled water required, using $C_1V_1 = C_2V_2$,

where:

C_1 = original concentration of stock solution,

V_1 = volume of stock required,

C_2 = concentration of final diluted solution,

V_2 = final total volume of diluted solution

Table 4.2 Volumes of stock and water used to produce diluted standard solutions

Sample number	Concentration of $K_2Cr_2O_7$ (mM)	Stock (ml)	Distilled water (ml)
1	0	0	2.0
2	0.25	0.4	1.6
3	0.5	0.8	1.2
4	0.75	1.2	0.8
5	1.0	1.6	0.4
6	1.25	2.0	0.0

Table 4.3 Absorbances of the standard and unknown solutions

Sample tube	Concentration (mM)	Absorbance at $\lambda 420$ nm	
		1st reading	2nd reading
1	0.00	0.000	0.000
2	0.25	0.205	0.150
3	0.5	0.438	0.325
4	0.75	0.563	0.524
5	1.0	0.725	0.675
6	1.25	0.941	0.900
7	unknown	0.450	0.447

Step 4: Using the blank (Sample 1) to set the spectrophotometer absorbance to 0, measure the absorbance for sample tubes 1 to 6 plus the unknown sample tube (Sample 7) twice, making sure you check calibration with blank before each diluted sample solution.

Step 5: Plot a graph of absorbance on the vertical axis and concentration on the horizontal axis, using the data in Table 4.3 below. Then plot the best-fit straight line (see Figure 4.5) through the data points.

Note: blanks and reagent blanks (see below for difference) do not necessarily show zero absorbance.

- Absorbance should increase proportionally with concentration provided the relationship is linear. If this is not apparent, check the blank absorbance, the cuvettes, and your sample queue which should show the same volume for all samples, and increasing intensity of colour with increasing concentration.

- If the solution is concentrated and the absorbance exceeds the value of 1.000, the relationship between absorbance and concentration may no longer be linear, which means dilution of the sample is necessary so that the absorbance values fit in the linear portion. Do not discard your samples before constructing the linear graph.

    e.g. Abs = 1.890 does not fit in the linear curve.

- Dilute the sample 5 times (1 ml sample + 4 ml distilled water, or a similar proportion in a smaller volume, i.e. 0.5 ml sample + 2.5 ml distilled water)
- Read the absorbance, and if the value is in the linear portion of the calibration curve, determine the concentration from the curve.

    Remember to multiply by the dilution factor of 5 in order to get the final concentration.

Step 6: Measure the absorbance of the unknown solution (blank first, then the unknown) and use the value to read the concentration from the standard curve (see figure 4.5). The absorbance of the unknown solution was 0.450, so therefore the concentration of the unknown solution is 0.625 mM.

Example 4.3

To make 2 ml (V_2) of $K_2Cr_2O_7$ at a concentration of 0.5 mM (C_2):

$$1.25 \times V_1 = 0.5 \times 2$$

$$V_1 = (0.5 \times 2)/1.25$$

$$V_1 = 0.8 \, ml$$

So, Standard 3 would use 0.8 ml of stock and 1.2 ml of distilled water to make the concentration of 0.5 mM.

4.2.4 Blanks and reagent blanks

The *blank* is the solvent (usually water) used in preparing the standard solutions. It is also called the *sample blank* or *reference blank*. However, in experiments where reagents have been added to the standards and samples, or where the samples have undergone a process (e.g. heating, boiling, incubating at certain temperatures, etc.), a *reagent blank* is prepared. This is a blank where the solvent (usually water) is treated exactly like the samples and standards, but does not contain the substance being analyzed. If the reagents or the treatment gives colour to the reagent blank, the absorbance value will be different from a sample or reference blank.

Thus, when drawing the calibration or standard curve, you will notice that the line of regression will not pass through the origin as it does with a sample or reference blank. In such cases, do not force the line through the origin unless you correct for the absorbance introduced by the reagents.

You can calculate *corrected absorbance* by subtracting the reading by the reagent blank from all the readings by the samples and standards. In such cases, the line of best fit is expected to pass through the origin.

Having found the unknown *concentration* from the standard curve, you may be asked to calculate the *amount* of the substance.

Example 4.4

Find the amount of a NaCl solution that has a concentration of 20 mg/L, with a total volume of 5 ml.

Remember, concentration is defined as mass/volume for liquids (e.g. mg/l), whereas amount is the total of the atomic masses of the elements in solution (e.g. in g, mg or mol).

Atomic mass of $NaCl$ = mass of Na + mass of Cl = $23 + 35.5 = 58.5$ g

Concentration = 20 mg/L

But the volume is not a litre; it is only 5 ml or $\frac{5}{1000}$ l.

Multiply $\frac{5}{1000}$ by the atomic mass 58.5.

Amount of $NaCl = \frac{5}{1000} \times 58.5 = 0.294$ g.

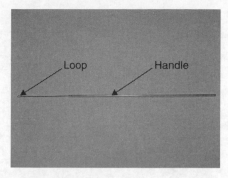

Figure 4.6 Inoculating loop

Figure 4.7 Bunsen burner

4.3 Aseptic techniques

Sterile or *aseptic* techniques are procedures used to prevent contamination of microbial cultures and the laboratory and surroundings, as well as contamination of yourself and others. Sterile techniques are used for laboratory preparation, culturing and transferring cultures, streaking plates to isolate and purify strains, and for cleanup and disposal.

4.3.1 Aseptic microbiological practice

Common apparatus used in aseptic microbiological work include Inoculating loops (Figure 4.6), Bunsen burners (Figure 4.7) and Petri dishes (Figure 4.8).

COMMON TECHNIQUES AND EQUIPMENT

Figure 4.8 Petri dish

Inoculating cultures

Step 1: The work area should be wiped with a disinfectant such as alcohol or bleach, and hands should be washed thoroughly with soap and water.

Step 2: Light the Bunsen burner. The 15–20 cm radius of the flame is effectively the sterile zone. The openings of test tubes and flasks must be flamed during the transfer of media or cultures. Heating the container directs air convection currents upward and away from the opening, momentarily preventing airborne contaminants from entering. It is essential to perform the transfer quickly, before the opening cools. Hold test tubes at a 45° angle in your left hand if you are right-handed, or in your right hand if you are left-handed. Use your other hand to hold the loop and to remove and hold the cap with your index finger.

Step 3: Flame the top of the tube. Take out the inoculum (the culture you are transferring), re-flame the opening and replace the cap.

Step 4: When working with Petri dishes, always hold the lid over the plate to prevent contaminants from landing on the surface of the agar. Remember to place the lid back on as soon as possible. When you lift the lid, hold it directly over the plate. To avoid contamination, do not place the lid on the bench top.

70

Step 5: When finished, turn off the burner and again wipe your work area with a disinfectant and wash your hands.

Step 6: Dispose of microbial cultures in a safe manner (either autoclaving all labware or soaking it in a bleach solution before disposal).

Streaking a plate

The purpose of streaking agar plates is either:

(i) to generate individual microbial colonies; or
(ii) to isolate individual microbial species from a mixture of microbes.

Streaking a plate

Step 1: The work area should be cleaned with a disinfectant such as bleach or ethanol.

Step 2: Label the plate on the bottom of the smaller half of the Petri dish. Make sure the date, the type of organism plated and your initials are clearly written along the edge so as not to obscure viewing the culture as it grows.

Step 3: Remove the top of the Petri dish and hold it in your left hand face down.

Step 4: Dip the sterile loop into the liquid culture.

Step 5: Decide the streak pattern to use for your plate. First touch the agar plate gently on the surface and, without lifting the loop from the surface, move it gently in a zigzag manner. If you dig into the agar, the cells will be deposited in a less aerobic environment, which reduces growth.

Step 6: Remove and flame the loop to sterilize it, then let it cool. Then perform another streak pattern. Using a new sterile loop for each streak is important in order to spread the cells enough to get individual colonies. It is important *not* to re-dip the loop in the original culture of bacteria after each flaming.

Step 7: Place the cover on the Petri plate and let the plate sit on the lab bench for about 20 minutes to allow the cells to adhere to the agar surface (Fig. 4.9a).

Step 8: Now, turn the Petri plate upside down to prevent condensation from dripping onto the agar surface and possibly causing contamination. The inverted plates will now be placed in

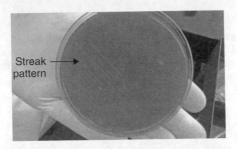

Figure 4.9a Streak plate

airtight plastic boxes or bags and incubated at whatever temperatures are directed for the method. The boxes or bags will keep the plates from drying out during the incubation time. The plates should be checked after two hours to ensure that there is no condensation on the agar surface or on the Petri plate lid. If there is condensation, put the lid on at a slant and allow the plate to dry out for 10–15 minutes. Then replace the lid fully.

Step 9: When the incubation period is finished, use a hand lens or stereomicroscope to examine the plate for colonies (Figure 4.9b). Continue observations of colony development and growth over several days. Note any apparent differences between colonies (e.gs colour, colony shape, colony edges, etc.).

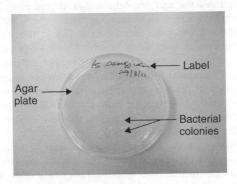

Figure 4.9b Plate with colonies

4.4 Disinfectants

Disinfectants are chemicals that destroy pathogenic bacteria from inanimate and skin surfaces.

4.4.1 Alcohols

Alcohols are used as they dehydrate cells, disrupt membranes and cause coagulation of proteins. Ethanol (80% v/v ethyl alcohol) or 2-propanol (60–70% v/v isopropyl alcohol) solutions are used to disinfect skin and decontaminate clean surfaces. Pure alcohols are not as effective as aqueous solutions. Other properties of alcohols as disinfectant include:

- Effective against fungi, vegetative bacteria, *Mycobacterium* species and some lipid-containing viruses.
- Generally not effective against spores, but methyl alcohol can kill fungal spores.
- Most effective at 70% in water. Prolonged contact time (5–10 minutes of wet contact) is required to be effective.
- May swell rubber or harden plastics.
- Do not use near flames due to flammability.
- **Disadvantages:** skin irritant, volatile (evaporates rapidly), inflammable.

4.4.2 Chlorine compounds

Chlorine based compounds are low to medium level disinfectants. Commonly used as bleach (sodium hypochlorite). Chlorine is usually used in concentrations of 1:10 (10%) or 1:100 (1%) for blood spills. Chorine bleach should never be mixed with other chemicals, as they can react to form toxic gases. Other properties include:

- Effective against a wide variety of microorganisms (bacteria and viruses). Use at 0.1% as a general disinfectant.
- Less suitable in the presence of organic matter such as blood.
- Effective between a pH range of 6–8.
- Strength decreases on standing as it oxidizes (make fresh solutions daily).
- High concentrations corrode metal surfaces such as steel, and can also bleach and damage fabrics.
- Chlorine can also damage rubber and some plastics.

4.4.3 Iodine

Iodine is used in aqueous or alcoholic solution. Its vapour is highly toxic and is absorbed through the skin. It is usually diluted to 1% w/v free iodine, optimum pH neutral to acid. Dilutions need to be prepared daily. Other properties include:

- Rapidly effective against most microorganisms.
- Most commonly used for skin disinfection and decontaminating clean surfaces. Dilute in alcohol for washing hands, or use as a sporicide.
- Not suitable in the presence of organic matter.
- Stains skin and may cause irritation.
- Decomposes when heated above 40°C.
- Do not use on aluminium or copper.

5

Microscopy and Histology

Learning outcomes

- To use a microscope, to observe and measure the size of a specimen.
- To count cells.
- To stain and mount a slide using haematoxylin and eosin.

5.1 Light microscopy

The light microscope is used to magnify or enlarge an object. It can also be used to measure the object size using special measuring scales called *eyepiece graticule* and *stage micrometer*. Normal eye resolution (the sharpness of the image) is approximately 0.1 mm, but the light microscope can increase this a thousand fold to 0.1 µm. Details observed under the microscope will also depend on contrast in image, i.e. the difference between light and dark areas.

There are different types of light microscope, such as monocular (one eyepiece), binocular (two eyepieces), compound and stereoscopic. Which particular one is used depends on the procedure performed (e.g. the stereoscopic microscope is used for dissections, since large specimens can be observed in a three-dimensional image).

The light microscope most commonly used in the lab is the bright-field compound microscope (Figure 5.1).

The microscope consists of:

- a movable stage which supports and holds the microscope slide in position – this comprises a mechanical stage and specimen holder clips;
- a lens system which concentrates light on the specimen – light source and condenser lens;
- an adjustable light source – brightness control and iris diaphragm to vary the amount of light;

Essential Laboratory Skills for Biosciences, First Edition. M.S. Meah and E. Kebede-Westhead.
© 2012 John Wiley & Sons, Ltd. Published 2012 by John Wiley & Sons, Ltd.

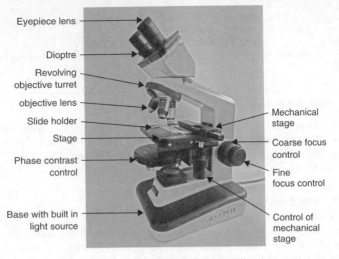

Eyepiece lens

Dioptre

Revolving objective turret

objective lens

Slide holder

Stage

Phase contrast control

Base with built in light source

Mechanical stage

Coarse focus control

Fine focus control

Control of mechanical stage

Figure 5.1 *Binocular light microscope*

- a mechanism for focusing the specimen by adjusting the distance between the lens system and the specimen – both course and fine focus;
- an imaging system to magnify and show detail – eyepiece lens and objective lens.

5.1.1 Lenses and magnification

Magnification is produced by two groups of lenses: the objectives, near the specimen (Figure 5.2a), and the ocular or eyepiece lens (Figure 5.2b) that have a magnifying power of $10\times$ (pronounced '10 times'). There are four to five objectives attached to a revolving turret or nosepiece: usually $4\times$, $10\times$, $20\times$, $40\times$ and $100\times$ (Figure 5.2c). The highest power is achieved with the oil immersion objective of $100\times$ which gives a total magnification of $1000\times$ with the magnification of the ocular lens ($10\times$).

The total magnification of the specimen is simply the multiplication of ocular power ($10\times$) and the objective power, e.g. If we used the $40\times$ objective lens, our total magnification would be $40 \times 10 = 400$.

Total magnification = power of ocular lens × power of objective

Figure 5.2a *Objective lens of microscope*

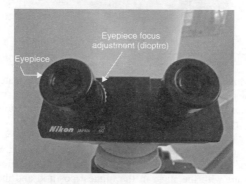

Figure 5.2b *Ocular lens of microscope (eyepieces)*

5.1.2 Using the microscope
The procedure for using the microscope is as follows:

Using the Microscope

Step 1: Switch on using the mains switch and adjust the brightness with the light intensity knob.

Step 2: Place the specimen slide on the stage using the specimen holder (the specimen side of the slide should be facing up). Make sure the slide is held securely by the metal clip (Figure 5.2c).

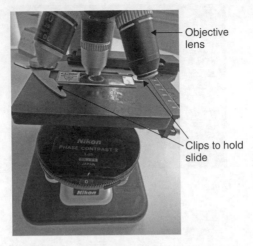

Figure 5.2c *Stage and clips to hold slide*

Step 3: Move the slide, using the knobs, until the material to be observed is illuminated by the light source. The x-axis knob moves the slide horizontally and the y-axis knob moves it vertically in the anterior-posterior direction.

Step 4: Select the $10\times$ objective using the revolving nosepiece or turret. Always begin with the lowest power objective to locate the specimen, then switch to a higher magnification.

Step 5: Start with the stage in the highest position and then bring it toward the objective, using the coarse focus adjustment knob. Note: Moving the stage up toward the objective without looking by eye can result in damage to the slide or objective.

Step 6: Bring the specimen in focus using the fine focus adjustment knob.

 Precautions

Always check the lens for dust or oil which might affect the viewing. Only use dry lens tissue; do not use any other type of paper or cloth, as these will scratch the delicate surface of the lens.

Optimizing the magnification and microscopic view

The microscopic view can be optimized by adjusting the ocular lens and the light:

- Adjust the interpupillar distance: while looking through the eyepieces, adjust for binocular vision (move each eyepiece apart) until the left and right fields of view coincide. If adjusted properly, you will be less likely to squint your eyes or look through one eye.
- Focus adjustment: Close your left eye while looking through the right eyepiece with your right eye and rotate the coarse and fine adjustment knobs to bring the specimen into focus. Then close the right eye and, while looking through the left eyepiece with your left eye, turn the dioptre adjustment ring on the left eyepiece (see Figure 5.3).
- Adjust the eyecups if necessary. If you wear spectacles, use the eyecups in the normal, folded-down position (to prevent your glasses scratching the eyepieces). If you do not wear glasses, extend the folded eyecups, pulling them towards your eyes.
- Light adjustment is made with the auxiliary lens centering knob and aperture and field iris diaphragms.

5.1.3 Phase contrast microscope

The advantage of the phase contrast microscope is that it can be used to observe living unstained cells. It is a light microscope which has two parts added to increase the contrast of transparent specimens without the use of the stains which would normally be used to show contrast.

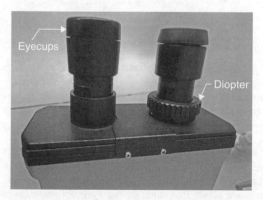

Figure 5.3 Adjustment of ocular lens

The first part of the phase contrast microscope is called a *phase annulus* (a black painted plate with holes) which creates separate light rays that are focused on the specimen. The second part is called the *phase plate* (a curved lens surface containing areas which absorb light). It works on the principle that some light rays passing through a specimen are diffracted by the dense part of the cell, while others go straight through the light part of the cell. The diffracted light is slowed by the phase plate, causing a difference in phase compared to the direct light, and this causes contrast (bright and dark areas) in the final image of the cell.

5.1.4 Using the light microscope under high power (100×) and oil immersion

This technique is often used to look at cell division stages (e.g mitosis or mieosis), living cells (e.g cheek epithelial cells), or very small cells such as blood cells or bacteria.

Using Oil immersion to view specimens

Step 1: Using the 10× objective, first locate the region to magnify and focus on the specimen.

Step 2: Place 1 drop of immersion oil on top of the cover glass.

Step 3: Rotate the 100× objective into the oil and light path.

Step 4: Use the fine adjustment knob to focus.

Step 5: Open the condenser diaphragm.

Step 6: Adjust the light intensity.

 Precaution

Avoid getting oil on other objective lenses. Since the magnification is 1000×, then any slight movement of the specimen or the stage is also magnified. Use the fine focus control and move the stage carefully. Make sure to clean the objective with lens paper after use.

5.1.5 Measurement of specimen size under the microscope

To measure the size of small objects under the microscope, we use two scales – eyepiece or ocular micrometer and stage micrometer. The stage micrometer scale is used to calibrate the eyepiece scale in micrometers, which is then used to calculate the size of specimens.

Equipment

The *ocular micrometer* (*eyepiece reticle*) consists of a small glass disc with a microscopic scale engraved across the diameter (see Figure 5.4). A typical eyepiece reticle would be a 5 mm or 10 cm linear scale with 50 or 100 divisions respectively. The disc is placed into the microscope eyepiece.

To use the eyepiece reticle for accurate measurements, it is necessary to calibrate the eyepiece reticle against a *stage micrometer*, a microscope slide with a scale etched upon its surface, usually 1 or 2 mm long, in 0.1 and 0.01 mm divisions (Figure 5.5). The stage micrometer is placed directly on the stage of the microscope and brought into focus.

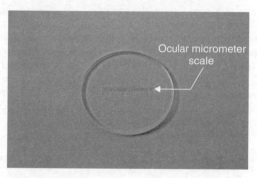

Figure 5.4 Ocular micrometer

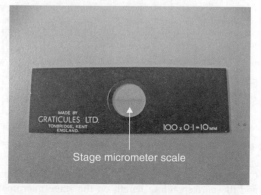

Figure 5.5 Stage micrometer

Calibration of the ocular micrometer

The ocular micrometer has to be calibrated for each objective lens used.

Calibration of the ocular micrometer

Step 1: Remove the eyepiece from the microscope and unscrew the lens at the top.

Step 2: Place the ocular micrometer in the eyepiece, screw the top lens back in place and replace the eyepiece.

Step 3: Place the stage micrometer on the stage with the 10× objective in place. Centre the marks of the slide in the field of view and focus (Figure 5.6 – the smaller parallel lines are the ocular micrometer scale with the numbers from 0-80 and the thicker longer parallel lines are the stage micrometer scale).

Step 4: Turn the eyepiece to place the scales in a parallel position and adjust the stage micrometer until the starting lines of both scales overlap completely. Figure 5.6 – in this figure we can see that 85 ocular divisions are equivalent to 28 divisions on the stage micrometer.

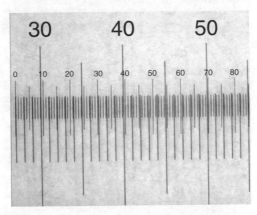

Figure 5.6 *Overlaying of ocular and stage micrometers*

Step 5: Count the number of divisions on the ocular micrometer and the corresponding length on the stage micrometer scale.

Notice that the scale for the stage micrometer starts at 27 and not 0; hence $55 - 27 = 28$. The eyepiece micrometer measures $0-85$, and dividing $28/85$ ($= 0.329$) tells us that one unit on the eyepiece is equal to 0.329 units on the stage micrometer. One division on the stage micrometer equals 0.01 mm, hence equals 0.00329 mm or 3.29 micrometer.

Step 6: Once you calculate the actual length of one division on the eyepiece micrometer you can use this value as the conversion factor for all measurements with the same objective and microscope set-up (i.e size of specimen = number of eyepiece divisions x conversion factor). The conversion factor for the given example above will be 1 ocular division = 3.29 μm. Following this calibration, all measurements with the ocular micrometer can be recorded and later converted to actual measurements in μm by multiplying with the conversion factor.

Repeat the same procedure with the other objective lenses if you are using different magnifications.

Example 5.1

Worked example:

If 1 division on the stage micrometer = 0.01 mm

and If 61 ocular divisions = 20 stage divisions

then 1 ocular division = 20/61 stage divisions,

so 1 ocular division = $(20/61) \times 0.01$ mm

$= 0.003278$ mm

$= 3.278$ μm
(since 1 mm = 1000 μm)

Measuring size under the microscope

Measuring size of specimen

Step 1: Place the specimen on the microscope stage and focus on the object to be measured.

Step 2: Superimpose the ocular micrometer scale and measure the dimensions of the object, i.e., count the number of divisions.

Step 3: Multiply the number of scale divisions by the conversion factor for the micrometer to give the actual dimension in micrometres.

5.2 Slide preparation

The study of the microscopic structure of organisms by observing stained cells under the microscope is called *histology*. Before tissues can be observed under the microscope, they have to undergo tissue processing (physical and chemical stages) in order to:

- preserve the tissue (using fixatives),
- produce thin sections (e.g. 3–5 μm thickness) and to allow light to penetrate, and
- show contrast (differences within the tissue) by staining using chemicals / dyes.

The result of tissue processing is first to produce a wax-covered unstained slide (Figure 5.7), which is then stained and finally mounted on a microscope slide to make a permanent slide. This can then be examined under the microscope.

There is a wide variety of stains in use, both natural and synthetic. Specific stains are used to show particular tissue structures, for example:

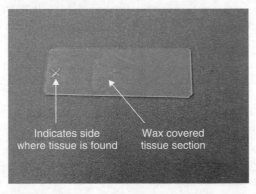

Figure 5.7 Unstained slide with wax covered section

- Periodic acid-Schiff shows carbohydrates.
- Masson's trichrome shows connective tissue (collagen, elastin).
- Van Gieson's stain shows connective tissue.
- Haematoxylin shows nuclei.
- Eosin shows cytoplasm.

The most popular stain used in histology is haematoxylin and eosin. Haematoxylin is a natural product (derived from the *Haematoxylum campechianum* tree) which is a basic dye that binds to nuclei to give a blue/purple colour. Eosin is made of many dyes derived from fluorescein (a fluorescent dye). It is an acidic dye that binds to substances in the cytoplasm to produce shades of pink and red.

There are many procedures for staining a tissue. Tissues are either stained slowly using a progressive method (keep adding dyes until the right colours are produced, as determined by checking frequently under the microscope), or by a faster method (2–3 hours) using a regressive method (overstain the tissue first and then remove the excess stain).

The procedure shown below is a popular regressive staining method using haematoxylin and eosin. Please note that there will be variations in the types of haematoxylin and eosin used, procedures used (e.g. levels of alcohol used in Step 2) and timing (e.g. time in Ehrlich's haematoxylin in Step 3) in different laboratories.

5.2.1 Haematoxylin and eosin staining

Haematoxylin and Eosin staining of an unstained slide

Step 1: Dewaxing

Place the wax-covered unstained slide section (Figure 5.7) upwards on a hot plate until the wax melts. The chemicals required for the staining process are shown arranged on the staining tray (Figure 5.8a). Then place the slide section in Histoclear solution for two minutes. The section should be transparent and no wax should be visible.

Step 2: Rehydrating tissue

- Place in 100% absolute alcohol for 2 minutes.
- Place in 90% alcohol for 2 minutes.
- Place in 70% alcohol for 2 minutes.

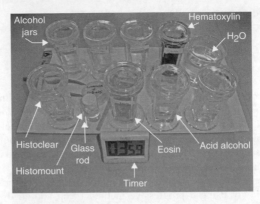

Figure 5.8a H & E staining tray

Step 3: Staining the nuclei

- Place in Ehrlich's haematoxylin for 30 minutes (Figure 5.8b).

Figure 5.8b Placing slide in Erlich's haematoxylin

- Remove the slide from the stain, holding tissue paper underneath it to catch the drips, then rinse for 30 seconds in clean water (tap water will do). The section should appear bluish-purple / dark blue.
- Place slide very briefly into acid alcohol solution for 3–4 seconds. The section will change from blue to red. This is to remove stain from the cytoplasm.

- Immediately place the slide into ammoniated (alkaline) alcohol solution and give it a wiggle until the section turns blue.

Step 4: Staining the cytoplasm

- Place section in 70% alcohol for 2 minutes.
- Place section in 90% alcohol for 2 minutes.
- Place in alcoholic eosin for 2 minutes.
- Place slide into 90% alcohol for 30 seconds.
- Place slide into absolute alcohol for 1 minutes.

Step 5: Clearing the section and preparing a permanent mount (Figure 5.8c)

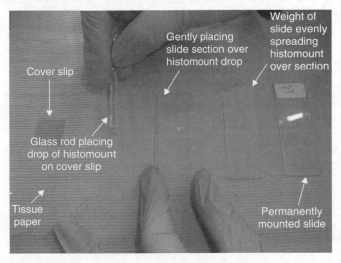

Figure 5.8c Making a permanent mount of stained tissue

- Place the slide into Histoclear for 2 minutes.
- While the slide is in the Histoclear, place a cover slip onto a piece of tissue and place one small drop of Histomount (a sticky clear resin) onto the centre of the cover slip (4–5 mm). Try to avoid bubbles by not twirling the rod in the Histomount. If you get bubbles or too much Histomount, use another cover slip and repeat.

- Remove the slide from the Histoclear and immediately place it section-side down onto the prepared cover slip, allowing the weight of the slide to spread the Histomount. Clear any excess Histomount, turn the slide over, let it dry, then label the slide with your name, the tissue and the stain used.

The slide should show the cell nuclei stained blue and other cellular components stained shades of pink and red.

5.2.2 Mounting
The final step in producing a slide is mounting, which can be permanent (Figure 5.8d) or non- permanent.

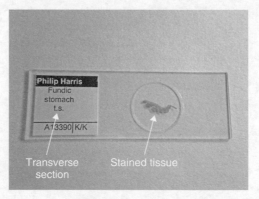

Figure 5.8d Permanent slide (stained)

Wet mount
This is used for fresh specimens (e.g. animal cells such as cheek cells, onion cells, protozoa, algae, etc.).

Non-permanent mounting of a specimen

Step 1: Put a small piece of dry specimen (e.g. onion epidermal cells) or 2–3 drops of liquid sample (e.g., with protozoa) on the microscope slide.

Step 2: For a dry specimen, add a drop of water or appropriate stain for transparent cells (e.g. Lugol's iodine for onion cells).

Step 3: Carefully place a clean cover slip over the specimen (to protect the objective lens), making sure that no air bubbles are trapped and there is no liquid spillage over the cover slip.

Step 4: Use the low-power objective (10×) to focus on the specimen and change to higher magnification depending on the specimen in view.

Dry mount
This is used for a non fresh specimen (e.g. a stained tissue slide – see Figure 5.8c).

Making a dry mount

Step 1: Add mounting medium (e.g. Histomount or Canada balsam) on the stained slide.

Step 2: Place a cover slip over the specimen, avoiding bubbles. Label the slide.

Step 3: Examine under low power.

5.3 Cell Counting

Quantitating the number of cells in a sample can be done using equipment such as a *flow cytometer*, or by microscopic count using a simpler device such as a *haemocytometer*, which is essentially a counting chamber. A good example of the use of a haemocytometer is the measurement of red and white blood cells.

5.3.1 Red blood cells (RBC) counting

The process of RBC counting involves the dilution of sample blood, since there are too many cells to count, and the removal of cells which are not red blood cells by using red blood cell diluting fluid.

The apparatus required includes a haemocytometer slide, a microscope, RBC diluting fluid, micropipettes and sample tubes. The haemocytometer is a glass slide (Neubauer) upon which are etched the grids used in cell measurement (Figure 5.9).

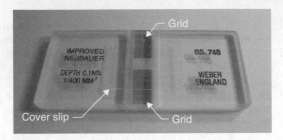

Figure 5.9 *Haemocytometer*

Protocol

Using a Haemocytometer

Step 1: Examine the glass slide by eye; you should see two grids etched in the centre of each end rectangle of the slide. Use one grid for the experiment.

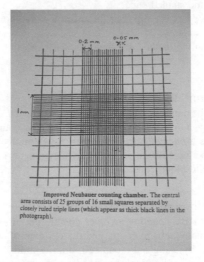

Improved Neubauer counting chamber. The central area consists of 25 groups of 16 small squares separated by closely ruled triple lines (which appear as thick black lines in the photograph).

Figure 5.10 *Haemocytometer grid with dimensions*

Step 2: Note the dimensions of the grid (Figure 5.10). It consists of nine squares of side 1 mm. The central square is split into 25

90

squares of side 0.2 mm. Each of these is split into 16 squares of side 0.05 mm. Typically, five 0.2 mm squares are used for the cell counting.

Step 3: Pipette 4 ml of RBC diluting fluid into a clean sample tube. Add 20 µl of blood (e.g. rat or guinea pig) and rinse the pipette with the diluting fluid. Gently agitate the contents of the sample tube.

Step 4: Moisten, using distilled water, the edges of the slide adjacent to the counting chamber. Place the cover slip firmly on the moistened edges so that the grid is covered.

Step 5: Using a micropipette, draw up some diluted blood, then touch the pipette tip to the junction between the slide and cover slip and squeeze out gently some diluted blood to cover the counting grid.

Step 6: Using high power (40× objective), count the RBC in five groups of 16 tiny squares (sides = 0.2 mm). Avoid counting cells more than once.

Step 7: Count the total number of cells in the 80 squares, and calculate the total number of RBC in 1 µl, using the formula: Number of RBC in 1 µl = total RBC counted × dilution factor/ volume of diluted blood where:

$$\text{dilution factor} = \text{final volume/original volume}$$

$$\text{volume of diluted blood} = \text{area of square} \times \text{depth}$$

$$\times \text{number of squares}$$

Note: 1 ml = 1000 µl; 1 mm^3 = 1 µl; depth = 0.1 mm.

6

Cardiorespiratory Measurements

Learning outcomes

- To measure heart rate.
- To measure blood pressure by manual and automatic methods.
- To obtain a finger-prick sample of blood.
- To perform a blood smear.
- To measure lung function using dry and wet spirometers.

6.1 Techniques to investigate cardiovascular function

The cardiovascular system consists of the heart pumping blood to the blood vessels of the circulatory system. Its function can be investigated by measuring heart rate, blood pressure, and parameters in blood samples (e.g. haemoglobin, red blood cells, glucose and lactate).

6.1.1 Measurement of heart rate

The *heart rate* is the number of beats per minute (bpm) that the heart undergoes. Typical heart rates for men and women is between 60–80 bpm at rest. Very fit endurance athletes can have heart rates of 28–40 bpm at rest, but poorly trained, sedentary individuals could have heart rates of 100 or more. The most accurate method of measuring heart rate is recording the electrical activity of the heart on an electrocardiogram (ECG) using an Electrocardiograph machine.

A simple method of estimating heart rate is to measure the *pulse*, the pressure wave created in the artery with each beat of the heart. The pulse can be felt at arteries near the surface of the body such as the radial artery in the lower arm or the carotid artery in the neck.

Essential Laboratory Skills for Biosciences, First Edition. M.S. Meah and E. Kebede-Westhead.
© 2012 John Wiley & Sons, Ltd. Published 2012 by John Wiley & Sons, Ltd.

Measurement of Pulse Rate using palpation

Step 1: First palpate for the pulse at the radial artery, located near the wrist on the thumb side.

Step 2: Place three fingers (not the thumb) over the location of the radial artery.

Step 3: Count the number of pulse beats in 15 seconds, starting at zero.

Step 4: Multiply the number of beats in 15 seconds by 4. This gives the heart rate in beats per minute.

Pulse rate is a good estimate provided the blood flow to the artery is normal. Pulse rate can also be measured by devices such as Pulse Oximeters and automatic blood pressure machines.

6.1.2 Measurement of blood pressure

Blood pressure is the pressure exerted by blood against the vessel walls. Arterial blood pressure is the most important. There are two major components:

- *Systolic pressure* (SP) is the *highest* pressure in the artery produced by the systolic contraction of the heart. Normal range $= 100 - 140$ mm Hg.
- *Diastolic pressure* (DP) is the *lowest* pressure in the artery, produced during diastole. Normal range $= 60 - 88$ mm Hg. Blood pressure is normally expressed as SP/DP.

Blood pressure can be measured by both manual and automatic methods.

Manual method

The principle of the manual method is auscultation (listening for sounds). The sounds are called Korotkoff sounds and are created by compressing and then relaxing the tissues around an artery. The traditional manual method of measuring blood pressure is by using a *sphygmomanometer* (Figure 6.1a) and a stethoscope, and measuring the blood pressure from the brachial artery of the upper left arm. The sphygmomanometer consists of an inflatable rubber bag (cuff) to squeeze the tissue surrounding the artery, a rubber bulb with valve to inflate and deflate the cuff, and a mercury or non-mercury manometer to measure pressure. The stethoscope (Figure 6.1b) is used for listening for sounds in the artery.

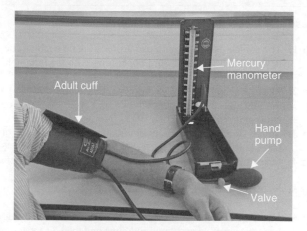

Figure 6.1a Sphygmomanometer

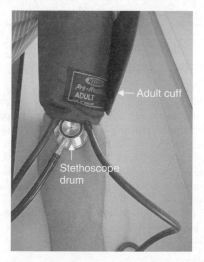

Figure 6.1b Sphygmomanometer and stethoscope

Measuring blood pressure manually

Step 1: The subject is seated with one arm resting on a table. The cuff is wrapped around the bare upper arm, with the inflatable bag over the inside of the arm.

Step 2: Place the bell of the stethoscope below the cuff and over the brachial artery (at the elbow), and close the valve on the hand pump.

Step 3: Slowly inflate the cuff by contraction and relaxation of the hand bulb to 160 mm Hg.

Step 4: Open the valve and lower the pressure in the cuff (e.g. between 1–3 mm Hg/sec).

Step 5: Note the pressure at the *first* sound you hear – this is the *systolic pressure*.

Step 6: Continue to release the pressure; when the sound suddenly becomes muffled and low in volume and then disappear, again note the pressure – this point is called the *diastolic pressure*.

Step 7: Allow the subject to rest and repeat again. Do not leave an inflated cuff on for more than two minutes.

 Precautions

- Make sure the cuff size is appropriate (i.e. for adult, child), as an oversized cuff will give a lower blood pressure reading and an undersized one will give an apparently higher blood pressure.
- See that the subject is at rest before taking measurement.
- Do not let the subject see the mercury manometer readings. Many people are treated for high blood pressure based on measurements in surgeries where the individual was stressed but their blood pressure was actually normal; this is known as 'white coat hypertension'.
- Ensure that the manometer is at the level of the heart.
- Check that there are no leaks in the system.

Automatic method

The automatic sphygmomanometer (Figure 6.2) measures blood pressure using the oscillometric prinicple – changes in oscillations or pulsations created by compressing and relaxing the artery. It consists of a cuff (which is inflated and deflated automatically), a pressure measuring device, power supply and a digital display.

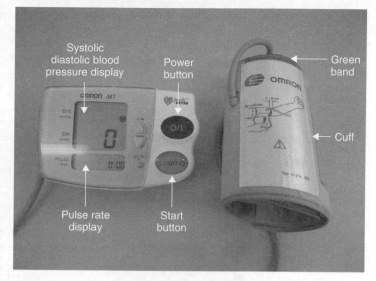

Figure 6.2 *Automatic blood pressure monitor*

The following procedure is used:

Measuring Blood Pressure by an automatic device

Step 1: Place the cuff on the bare left upper arm.

Step 2: Make sure that the green line on the edge of the cuff is just above the elbow.

Step 3: Press the start button and wait until the digital display shows the systolic pressure, diastolic pressure and pulse rate in the sitting posture.

Step 4: Note values and repeat after a rest period of one minute.

 Precautions

- See that the subject is at rest.
- Do not let the subject see the display on the blood pressure monitor.
- See that the monitor is at the level of the heart.
- Check that the monitor has been calibrated.

6.1.3 Obtaining a finger-prick sample of blood

Small quantities of blood (e.g. $20-30\,\mu l$) are often required to analyze substances such as blood glucose and lactate. It is common and simple to use finger-prick 'pens'. These contain a lancet which pierces the skin when the pen is fired.

Two examples of pens which are readily used include the 'Accu-Chek Softclix Pro' and the 'OneTouch UltraSoft'.

Using the Accu-Chek Softclix Pro pen to obtain blood

Step 1: Set the depth of lancet penetration (size $1-3\,mm$) by rotating the ring at the end of the pen (Figure 6.3).

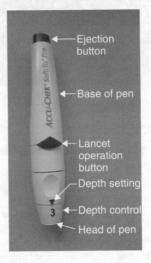

Figure 6.3 Finger prick pen

Step 2: Twist the end ring of the pen to open the aperture and insert the thin end of the lancet fully (Figure 6.4a, Figure 6.4b).

Step 3: Twist the handle of the lancet and remove handle (Figure 6.4c, Figure 6.4d).

Step 4: Gently place pen against side (Figure 6.4e) and end of finger (use third or fourth finger).

Step 5: Press the blue button to pierce the skin and collect blood.

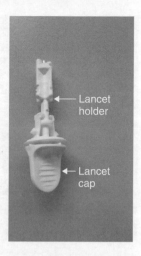

Figure 6.4a Lancet

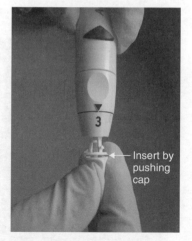

Figure 6.4b Pushing lancet into pen

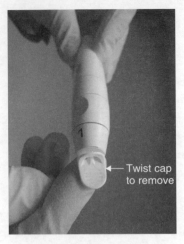

Figure 6.4c Removing base of lancet

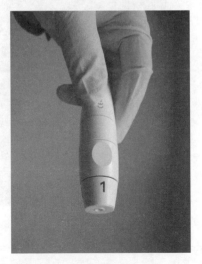

Figure 6.4d Pen containing inserted lancet

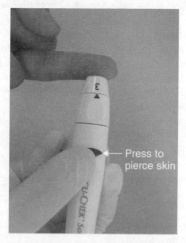

Figure 6.4e *Pen placed perpendicular to side of finger*

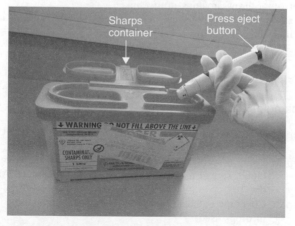

Figure 6.4f *Ejection of used lancet from pen*

Step 6: Press the blue knob at the base of the pen to eject lancet into 'sharps' container (Figure 6.4f).

Step 7: Cover the site of puncture with cotton or sterile plaster.

There are only small variations in the procedure if you use the OneTouch UltraSoft pen – mainly in the method of attaching the lancet.

6.1.4 Blood smears

A blood smear is useful to investigate the type and number of different white blood cells (leukocytes). White blood cells have different sizes and are nucleated. Six different types should be identifiable: granulocytes (neutrophils, eosinophils, basophils) and agranulocytes (small and large lymphocytes, monocytes).

The materials required are samples of blood, glass slides, a light microscope, sterile lancet or finger pricking device (see Section 6.1.3), Wright's stain and phosphate buffer solution.

 Precautions

- Only handle your own blood.
- Wear latex gloves.
- Only use a sterile lancet once.
- Do not throw anything in the sink.
- Leave waste solutions and items such as lancets, slides and capillary tubes in sharps containers.

Procedure

Making a Blood Smear

Step 1: A drop of blood obtained by finger puncture is placed on one end of a glass slide.

Step 2: Using a second slide held at approx 45° to the edge of the blood, spread the blood evenly over the first slide. Allow to dry.

Step 3: Cover the blood smear with Wright's stain (count the number of drops used).

Step 4: After two minutes, add an equal number of drops of buffer solution and blow to mix buffer and stain. Let the slide stand for four minutes. Flush the slide with tap water, then let it dry.

Step 5: Examine the smear under low power, then under oil immersion, and identify the different leukocytes. Ideally, count 100 white blood cells and express the types as percentages.

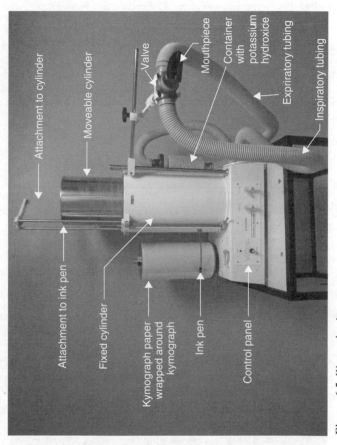

Figure 6.5 Wet seal spirometer

6.2 Techniques to investigate respiratory function

Measurements of respiratory function can be assessed by performing lung function tests during normal breathing and while performing artificial breathing manoeuvres. These tests include measuring volumes using *spirometer*s and measuring expired gases using *Douglas bags*. Two common spirometers used include the cylindrical *wet spirometer* and the dry *Vitalograph*.

6.2.1 Using a wet spirometer

The instrument (Figure 6.5) consists of two cylinders: a fixed cylinder containing water, and a movable aluminium cylinder inside the fixed cylinder which moves up and down with breathing movements (*inspiratory* and *expiratory*). These breathing movements are recorded using a pen on a rotating drum called a *kymograph* via a pulley system attached to the movable cylinder. The wet spirometer can also be used to measure metabolic rate.

Recording a spirogram

A spirogram is a recording of lung volumes whilst breathing on a spirometer. The procedure for its use is as follows:

Recording a Spirogram

Step 1: The movable cylinder is first filled with oxygen via the gas inlet tube (Figure 6.6a).

Figure 6.6a Controls of wet seal spirometer

Step 2: The tubing is checked to ensure that the inspiratory tube is coming from the moveable cylinder and the expiratory tube is going to the movable cylinder via the container of potassium hydroxide solution (to absorb the expired CO_2).

Step 3: A mouthpiece is attached to the valve. The subject puts on a nose clip and then closes their mouth around the mouthpiece.

Step 4: To connect the subject so that they are breathing from the spirometer, switch the lever from atmosphere to spirometer (Figure 6.6b).

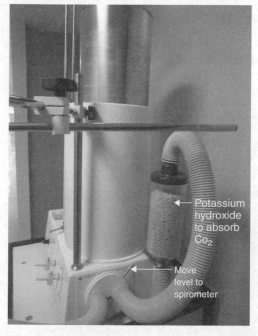

Figure 6.6b *Controls to switch between spirometer and atmosphere*

Step 5: Record normal breathing for 1 minute at a kymograph speed of 25 mm/min, then increase the speed to 2.5 mm/s and ask the subject to inhale maximally and then exhale, then go back to normal breathing.

Step 6: Using the vertical volume scale in ml and horizontal scales in mm/min or mm/s, calculate from the recording, breathing frequency (number of breaths per minute), minute ventilation (volume of air expired or inspired per minute) and lung volumes (tidal volume, inspiratory capacity, expiratory capacity, vital capacity).

6.2.2 Recording forced expiratory volumes using a dry spirometer

The Vitalograph (Figure 6.7) is a dry spirometer consisting of a bellows connected to a tube exposed to the air. Air entering the bellows via the tube causes the expansion of the bellows, and this is recorded via a metal pen on special pressure-sensitive paper placed on a movable tray. This paper has time in seconds on the horizontal axis and volume in litres on the vertical axis.

The Vitalograph is used to measure two lung volumes: forced vital capacity (FVC) and the volume expired in the first second (FEV_1) after a maximal inspiration. These volumes are used to assess normal lung function. The procedure is as follows:

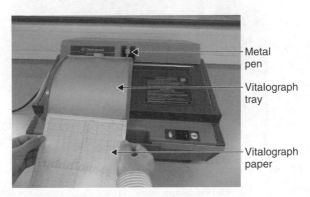

Metal pen
Vitalograph tray
Vitalograph paper

Figure 6.7 *Loading vitalograph chart paper onto movable tray*

CARDIORESPIRATORY MEASUREMENTS

Recording FEV$_1$ and FVC

Step 1: Attach the Vitalograph paper to the tray and check that the pen is on the zero position of the horizontal time axis (Figure 6.8a).

Step 2: Attach a disposable cardboard mouthpiece at the end of the tube.

Step 3: The subject inhales air maximally, then places their mouth around the end of the disposable mouthpiece (Figure 6.8a) and blows as fast and as hard as possible. Notice that the vitalograph pen has moved from the left to the right of the tray, recording a trace of the breathing manoeuvre (Figure 6.8b). Encourage the subject to perform the blow with the greatest effort.

Step 4: After a rest of 30 seconds, repeat the manoeuvre two more times.

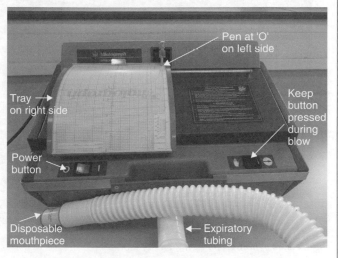

Figure 6.8a Vitalograph at the beginning of a blow

Step 5: From the vitalograph paper, record FEV$_1$ by moving along the time axis to the 1 second mark, then draw a line vertically until it crosses the recorded vitalograph trace. Read off the value in litres from the vertical axis (volume). The FVC is read from the vertical axis (volume) by drawing a line along the plateau of the vitalograph trace. The highest values are expressed as a percentage of predicted values based on age, gender, height and ethnicity, i.e.:

Percentage difference in FVC = experimental FVC
$\times$ 100/predicted FVC

Percentage difference in FEV_1 = experimental FEV_1
$\times$ 100/predicted FEV_1

If the percentage difference is 80 percent or greater, then the subject has normal lung function.

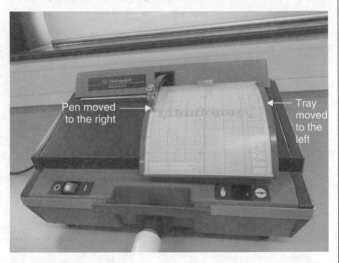

Figure 6.8b *Vitalograph at the end of a blow*

6.2.3 Recording forced expiratory volumes using an electronic spirometer

Electronic spirometers such as the MicroPlus (Figure 6.9) can also measure lung function. The MicroPlus uses a turbine flowhead to measure volumes and flows. Rotation of the turbine is proportional

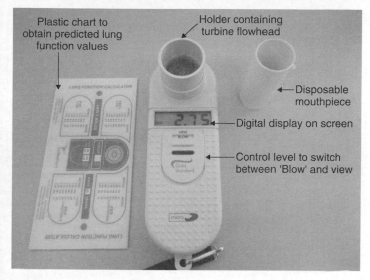

Figure 6.9 *Electronic spirometer (MicroPlus)*

to the airflow, and calculated values of FVC (Forced Vital Capacity), FEV_1 (Forced Expiratory Volume in the first second) and PEF (peak expiratory flow) are shown on the digital display.

 Precautions

Check calibration to make sure there are no leaks around the mouthpiece.

Recording FEV_1, FVC and PEF using an electronic spirometer

Step 1: Connect a disposable cardboard mouthpiece to the inlet on the spirometer.

Step 2: Switch the setting from off to 'Blow'; the display should show 0.00.

Step 3: While holding the spirometer vertically, take a maximal breath in and then blow into the mouthpiece as hard and fast as possible until you are unable to blow any more air out.

Step 3: Switch setting to 'view' and note the readings for FVC, FEV_1 and PEF.

Step 4: Repeat two further times and note the highest values of the lung function indices.

Step 5: Using the lung function calculator card (see Figure 6.9), compare your values for normality using the criteria of gender, height and age.

7

Recording and Presenting Data

Learning outcomes

- To know what to record in a laboratory book.
- To present data in tables.
- To present data in graphs.
- To describe data using descriptive statistics.

It is standard practice in laboratory practical work to provide students with a practical schedule, introducing the topic and itemizing the materials and methods of the practical. You should read all instructions carefully before proceeding with the tasks. Make notes of any changes in methods that may be given in the practical briefing by the demonstrator or tutor at the beginning of the practical.

7.1 Keeping a laboratory book

Using a bound A4 size laboratory notebook, record all data and observation in *ink* during your laboratory session; you should not do it afterwards from pieces of paper! You can leave some space for any further information you may want to add to your lab book before your next practical or experiment. Remember, the lab book is an active log of what you observe and what you do, measure and analyze.

- Write the date, title, and aim of the experiment.
- Refer to the method or procedure being used. If it is not to be handed in, stick your schedule into your lab book or write a brief description of the method.
- Identify and record the hazards and risks associated with any chemicals and equipment being used.
- Record data correctly and legibly, with detailed working of calculations. This will make it possible to see where mistakes have been made, such as calculations of molecular weights, dilutions or concentrations.
- Make sure you don't confuse similar symbols (e.g. M and m).

Essential Laboratory Skills for Biosciences, First Edition. M.S. Meah and E. Kebede-Westhead.
© 2012 John Wiley & Sons, Ltd. Published 2012 by John Wiley & Sons, Ltd.

- Only record data which has been measured; science is about observing, recording and analyzing real data, not trying to produce perfect results.
- Record numbers to an appropriate number of significant figures, and with the correct units, e.g. mg, g, mg l^{-1}, etc.
- Interpret data in the form of graphs, spectra, etc.
- Record conclusions.
- Identify any actions for future work.
- Make sure to record any changes or modifications you or the instructor made to the practical schedule.

Questions to ask about your data include:

- Does the data make sense? For example, is it within the expected range? With increasing concentration of a substance, absorbance is expected to increase proportionally – if it does not, then check the sample.
- Are there very high or low values in comparison to other values? Do not eliminate an anomaly immediately, but record it and consider its use during analysis. Can you give possible explanations to the anomalous results?

7.2 Presentation of data

Data collected from an observation, practical or experiment will contain a mixture of *values* and *variables* (independent and dependent) that need to be presented either in tabular or graphical form.

The *independent variable* is the variable controlled or manipulated in an experiment, while the *dependent variable* is the possible result or effect of the independent variable (see Table 7.1). In simple words, the independent variable may be the possible 'cause' and the dependent variable the 'effect'.

Independent variables may also be factors that the observer cannot manipulate – for example, environmental factors such as time, temperature, etc. In testing the effect of temperature on growth of bacteria in the soil, experiments can be conducted in the laboratory, where the temperature is the controlled independent variable. Field measurements of the same variables can be tested for relationship where temperature is still the independent variable, but not controlled as in the lab experiment.

7.3 Recording data in tables

All tables should have an appropriate title which describes the data, and this should be numbered and written above the table (see Table 7.2). Columns contain details of variables (independent and dependent) and their units, usually with data that need to be compared with different

Table 7.1 *Examples of independent and dependent variables*

Experiment	Independent variable	Dependent variable
Growth of bacterial cells with time	Time	Bacterial cell numbers
Effect of temperature on growth of bacterial cells	Temperature	Bacterial cell numbers
Relationship of concentration to absorbance on the spectrophotometer (i.e. standard curve)	Concentration	Absorbance

Table 7.2 *Glucose tolerance test for normal and diabetic persons*

Time (hours)	Plasma glucose (mg per 100 ml)	
	Normal	Diabetic
0.0	73	112
0.5	125	180
1.0	140	225
1.5	125	235
2.0	82	233
2.5	70	220
3.0	70	210
3.5	71	207

parameters (e.g. length, width of cells), while rows may contain different treatment or organisms, etc. (e.g. species 1, species 2, variable time). Numerical values in a column representing a parameter are generally presented to the same number of decimal places.

7.4 Presenting data in graphs

Graphs are used to describe a relationship between two variables, e.g. x and y. It is normal practice to identify the x-axis as the horizontal axis representing the independent variable, e.g. concentration. The y-axis is vertical and used to plot the dependent or response variable, e.g. absorbance of light. Each axis needs a label with the appropriate units of measurement. If mean values are plotted, then error bars (either standard

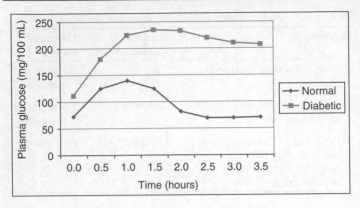

Figure 7.1 *Example of line graph*

error or standard deviation) should be shown. If there is a large gap from zero to the first data point, then it is common to show *broken axes*.

Graphs need a title that describes the presented data. The title should be numbered and written below the graph (see Figure 7.1). Do not use the word 'graph' in the title, but rather 'Figure' or 'Fig.'.

The most common graph types used to present data in the lab are line, bar and scatter graphs.

A *line* graph would be used if samples or measurements were taken at regular time intervals or with increasing/decreasing amounts (Figure 7.1), while *bar* graphs would be used if samples were not along a continuous variable such as increasing time or amount (Figure 7.2).

Where the relationship between the *x* and *y* needs to be drawn and the correlation tested, a *scatter* graph is used (Figure 7.3). The data points are plotted to show the spread of values along the two axes.

The line of best fit is drawn based on the relationship tested (Figure 7.4). In the lab, you will be drawing several calibration curves for many analytical and bioassay analyses.

The mathematical relationship most commonly used for calibration is the linear relationship or linear regression:

$$y = mx + c$$

where:

> *y* is the response, e.g. absorbance or signal (mV)
> *x* is the concentration of the working solution
> *m* is the slope or the gradient of the line graph
> *c* is the intercept of the line graph on the *y*-axis.

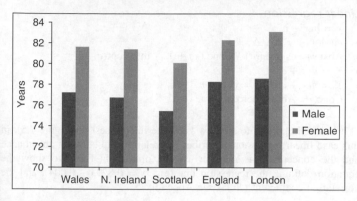

Figure 7.2 Example of bar graph

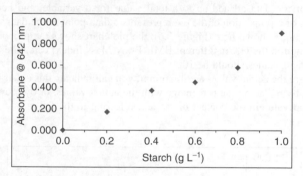

Figure 7.3 Example of scatter graph

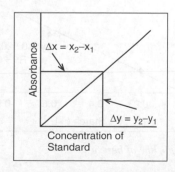

Figure 7.4 Drawing line of best fit using linear regression

Linear regression

- $y = mx + c$
- $m(\text{slope}) = \Delta y / \Delta x$
- $\text{Abs}(y) = \text{slope}\,(m)\,X\,\text{Conc.}(x) + Y - \text{intercept}(c)$
- $Y = mx + c$
- $Y = mx + c;\quad x = y - c/m$
- $\text{Conc.} = \text{Abs.} - \text{intercept/slope}$

It is also important to see how good the line of best fit (Figure 7.5), in this case linear regression, describes the relationship between absorbance and the concentration. You can use the options in Excel to show the equation defining the regression line (e.g. $y = 0.829x - 0.004$), and the correlation coefficient (R^2) – see Appendix 5.

Some types of data are best presented in a *pie chart*, where sections or parts of the data all add up to a total value for a variable, and hence a section or a proportion of the pie represents a data point in the data set. In the example shown here (Figure 7.6), the pie chart shows the proportions of women in the UK in different BMI (Body Mass Index) categories. The sum of all values should be 100%.

Where the actual values are important, you can indicate this either in the figure, by replacing the percentage with the actual values, or by providing the total value in the content of the text referring to the figure.

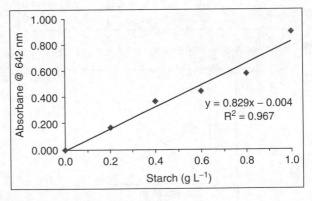

Figure 7.5 Testing the line of best fit

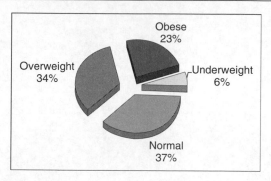

Figure 7.6 Pie chart

7.5 Describing data statistically

As well as being displayed in tables and graphs, data can also be described statistically by using descriptive statistics such as the number of data points (N), the average of the data points (mean), the standard deviation of the data points (SD) and the standard error of the mean (SEM). The meaning of these terms and formulae to calculate them are shown in Appendix 5.

Descriptive statistics can also be described graphically in figures and tables. All figures and tables should have titles and legends which can include details such as N, SD and SEM. In graphs, mean values are usually plotted with error bars of either SD or SEM. Significant differences between means are usually indicated by an asterisk next to the bar or data point.

7.5.1 Use of software to create tables and graphs, and perform statistics

Tables and graphs can be completed in laboratory books by hand, or are increasingly being produced using software such as Microsoft *Excel* or Microsoft *Word*. Other software packages you may come across include *Graphpad Prism* and *SPSS*.

Appendix 6 provides some details of producing tables and graphs using these software packages.

Recommended Reading

Ahern, H. (1992) *Introduction to Experimental Cell Biology*. Wm. C. Brown, Dubuque, IA.

Ashcroft, S. & Pereira, C. (2003) *Practical Statistics for the Biological Sciences*. Palgrave Macmillan, Suffolk.

Basics of Light Microscopy & Imaging, 2nd Edition. Available at: www.imaging-git.com/news/basics-light-microscopy-imaging (Accessed: 7th November 2011).

Bland, M. (1997) *An Introduction to Medical Statistics*. Oxford Medical Publications, Oxford.

Cann, A.J. (2003). Maths from Scratch for Biologists. John Wiley & Sons, Chichester. Chemlab. Available at: www.dartmouth.edu/~chemlab (Accessed: 8 November 2011).

Cook, D.J. (1998) *Cellular Pathology*. Butterworth Heinemann, Oxford.

Dean, J.R., Jones, A.M., Holmes, D., Reed, R., Weyers, J. & Jones, A. (2002) *Practical Skills in Chemistry*. Pearson Education, Harlow.

Ennos, R. (2000) *Statistical and Data Handling Skills in Biology*. Prentice Hall, Harlow.

Gas-liquid chromatography. Available at: www.chemguide.co.uk/analysis/chromatography/gas.html (Accessed: 18 July 2011).

Jones, A., Reed, R., Weyers, J. (2007) *Practical Skills in Biology*, 4th Edition. Prentice Hall, London.

Junqueira, L.C., Carneiro, J., Kelley, R.O. (1992) *Basic Histology*, 7th Edition. Appleton and Lange, Norwalk, CT.

Kaplan, A., Jack, R., Opheim, K.E., Toivola, B., Lyon, A.W. (1995) *Clinical Chemistry (Interpretation and Techniques)*, 4th Edition. Williams and Wilkins, Philadelphia, PA.

Essential Laboratory Skills for Biosciences, First Edition. M.S. Meah and E. Kebede-Westhead.
© 2012 John Wiley & Sons, Ltd. Published 2012 by John Wiley & Sons, Ltd.

Luxton, R. (1999) *Clinical Biochemistry*. Butterworth-Heinemann, Oxford.

Phoenix, D. (1997) *Introductory Mathematics for the Life Sciences*. Taylor & Francis, London.

Plummer, D.T. (1987) *An Introduction to Practical Biochemistry*. McGraw-Hill, London.

Pocock, G., Richards, C.D. (2009) *The Human Body*. Oxford University Press, Oxford.

Reed, R., Holmes, D., Weyers, J., Jones, A. (1998) *Practical skills in Biomolecular Sciences*. Addison Wesley Longman, Edinburgh.

Reed, R., Holmes, D., Weyers, J. & Jones, A. (2007) *Practical skills in the Biomolecular Sciences*, 3rd Edition. Benjamin Cummings, London.

Robyt, J.F. & White, B.J. (1987) *Biochemical Techniques Theory and Practice*. Brooks/Cole Publishing Company, Monterey, CA.

Selected Laboratory Methods. Available at: www.ruf.rice.edu/~bioslabs/methods/methods.htm (Accessed: 7 November 2011).

Tharp, G.D. (1993) *Experiments in Physiology*, 6th Edition. Macmillan Publishing Company, New York.

The Interactive Lab Primer. Available at: www.chem-ilp.net (Accessed: 8 November 2011).

Vodopich, D.S., Moore, R. (2008) *Biology Laboratory Manual*. McGraw Hill, New York.

Webster, J.G. (2004) *Bioinstrumentation*. John Wiley & Sons, New York.

Wilson, K. & Walker, J. (2010) *Principles and Techniques of Biochemistry and Molecular Biology*, 7th Edition. Cambridge University Press, Cambridge.

Young, B. & Heath, J.W. (2006) *Wheater's Functional Histology: A Text and Colour Atlas*, 5th Edition. Churchill Livingstone, Edinburgh.

Appendix 1: Rules for Powers

1. Powers can be positive numbers, negative numbers, decimals or fractions.

2. Before following the rules below, check that the bases are the same.

3. In the rules below, the base is denoted by 'Y' and the power is denoted by 'a' and 'b'.

   (i) $Y^a \times Y^b = Y^{a+b}$: in multiplying, add the powers.

   (ii) $Y^a \div Y^b = Y^{a-b}$: in dividing, subtract the powers.

   (iii) $Y^{-a} = 1/Y^a$: a negative power is equivalent to the reciprocal of the base raised to a positive power.

   (iv) $Y^a = 1/Y^{-a}$.

   (v) $Y^{a/b} = \sqrt[b]{Y^a}$ — a fraction power is equivalent to the 'b' root of Y raised to the power of 'a'.

   (vi) $Y^{1/2} = \sqrt{Y}$ = square root of Y.

   (vii) $Y^{2/3} = \sqrt[3]{Y^2}$ = cube root of Y squared.

   (viii) $Y^0 = 1$: any number raised to the power of zero is always 1.

Essential Laboratory Skills for Biosciences, First Edition. M.S. Meah and E. Kebede-Westhead.
© 2012 John Wiley & Sons, Ltd. Published 2012 by John Wiley & Sons, Ltd.

Appendix 2: Rules for Logarithms

1. The rules for logs are similar to the rules for powers. They hold true assuming all logs are to the same base, typically base 10.

2. Numbers less than 1 will have negative logs.

3. Negative numbers do not have logs.

4. If you know the log, then finding the antilog is the means to find the original number. If log = a, then $\text{antilog}_{10}a = 10^a$

   (i) $\text{Log}\,(a \times b) = \log a + \log b$

   (ii) $\text{Log}\,(a \div b) = \log a - \log b$

   (iii) $\text{Log}\,a^n = n \log a$

Essential Laboratory Skills for Biosciences, First Edition. M.S. Meah and E. Kebede-Westhead.
© 2012 John Wiley & Sons, Ltd. Published 2012 by John Wiley & Sons, Ltd.

Appendix 3: Factors to Consider When Making Solutions

Temperature

Water is at its highest density when its temperature is 4°C. 1000 ml of water at this temperature would weigh 1000 g. However, water expands and contracts with temperature changes, which causes changes in volume and therefore changes in its density. In practice, since these changes are small, we rarely measure the temperature when making solutions. Conventionally, solutions are made at a temperature of 20°C.

Solubility

Some substances have poor solubility in water. To remedy this, you may have to dissolve the substance in hot water in a beaker, and after it has cooled, add it to the volumetric flask. Hot water can not be used for substances that may be labile to heat (e.g., enzymes). So you have to read instructions carefully before you apply heat. Some organic substances need to be dissolved in ethanol before making the aqueous solution.

Mass of solute

When we make a molar solution, we use a mole of solute (i.e. the molecular weight of the solute in grams) dissolved in a little water, which we then make up to 1 litre. Thus the mass of the solute varies, and this dictates how much water needs to be added. A *molal* solution, however, is a solution in which the volume of water is 1000 ml and the amount of solute is one mole. Here the mass of solute varies, but the volume of water is kept constant.

Essential Laboratory Skills for Biosciences, First Edition. M.S. Meah and E. Kebede-Westhead.
© 2012 John Wiley & Sons, Ltd. Published 2012 by John Wiley & Sons, Ltd.

Appendix 4: Principle of Spectrophotometry

Spectrophotometry is used to determine the concentration of substances in solution by the amount of light energy they absorb.

The parts of a typical spectrophotometer consist of a light source (e.g tungsten lamp), a monochromator to separate light of a particular wavelength (e.g prism + slits), a cuvette (to hold the solution), photodetector (to convert light to electrical energy) and a readout device giving values for absorbance or transmittance (see Figure A4.1).

Transmitted light that passes through the sample is detected by a photodetector and measured to yield the transmittance or absorbance value (optical density) for the sample. Transmittance is expressed as a ratio, and is a measure of the light leaving (I) and the light entering (I_0) the cuvette. As concentration of substance increases, the transmittance decreases. Transmittance is inversely related to absorbance; hence absorbance increases with concentration.

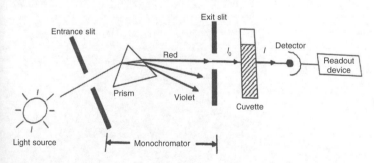

Figure A4.1 *Components of a spectrophotometer*

Essential Laboratory Skills for Biosciences, First Edition. M.S. Meah and E. Kebede-Westhead.
© 2012 John Wiley & Sons, Ltd. Published 2012 by John Wiley & Sons, Ltd.

APPENDIX 4: PRINCIPLE OF SPECTROPHOTOMETRY

The Beer-Lambert law states the relationship between absorbance and concentration, i.e. absorbance is directly proportional to concentration:

$$A = \sum lc$$

Where:

$\sum$ = constant called the *extinction coefficient* of the substance (units $M^{-1} \times cm^{-1}$)

l = the sample path length (width of cuvette usually 1 cm)

c = the molar concentration of the solution

Appendix 5: Descriptive Statistics and Formulae

After collecting numerical data from experiments, it is common to see how the data is distributed by drawing a frequency distribution graph (frequency of occurrence of data on the vertical axis and classes or groups of data on the horizontal axis).

Ideally we would be expecting a 'normal' (bell-shaped) distribution of the data, but this may not be the case, if for example, you only had a small sample. This distribution can be described by using descriptive statistics to indicate the central tendency and its variation around the central tendency (i.e. spread of data).

Central tendency

$$Mean \text{ or } average = \text{sum of } x/N,$$

where:

x = data value
N = total number of data

The *median* is the central or middle number of the data.
The *mode* is the most frequent value.

Variation of data

- The *range* is the largest value minus the smallest value.
- *Standard deviation* (SD) is a measure of the distribution of data around the mean.

   If the data is normally distributed, then SD tells us what proportion of the scores falls within certain limits, e.g. 95% between ±2 SD of the mean.

$$SD = \sqrt{[(\text{summation of } x - mean \text{ of } x)^2/(N-1)]}$$

$$= \sqrt{\left[\sum(x - mean)^2/(N-1)\right]}$$

- *Variance* is the square of the standard deviation (SD^2).

Essential Laboratory Skills for Biosciences, First Edition. M.S. Meah and E. Kebede-Westhead.
© 2012 John Wiley & Sons, Ltd. Published 2012 by John Wiley & Sons, Ltd.

- *Standard error of the mean* (SEM) is a measure of the accuracy of the mean.

$$SEM = SD/(\text{square root of } N)$$

95% confidence interval for the mean

$$(CI) = \text{mean} \pm 1.96\ SEM(\text{of population})$$

- *Confidence interval* (CI) gives the values between which there is 95% confidence of finding the population mean. If $N < 30$, then use the t distribution:

$$CI = \text{mean} \pm t_{value} \times SEM(\text{of sample}),$$

where

$$t_{value} = \text{value corresponding to}$$
$$p = 0.05 \text{ and degrees of freedom } (N - 1).$$

Coefficient of variation (%)

In repeated measurements, there is variation in the results. The size of this variation in relation to the size of the quantity being measured can give some indication of accuracy. A measure of relative variation is called the coefficient of variation (CV), which is expressed as a percentage. The higher the CV, the higher the variation and the lower the accuracy.

$$CV = (SD/\text{mean}) \times 100$$

Accuracy of a measurement

$$\text{Percentage error} = (\text{experimental mean value-true value})/$$
$$\text{true value} \times 100$$

Precision of an experiment can be expressed as the relative standard deviation (coefficient of variation) by calculating the standard deviation of the repeated measurements divided by the mean of the repeated measurements, i.e.:

$$\text{Percentage relative standard deviation} = \frac{\text{SD of repeated measurements}}{\text{Mean of repeated measurements}} \times 100$$

Correlation and r^2

r^2 (coefficient of determination) is a measure of the linear relationship between two variables, and has a value between 0 and 1. A value close to 1 indicates high correlation but does not mean one causes the other. For example, a value for r^2 of 0.9 suggests that 90 per cent of the variance of the x-axis variable can be explained by the variation of the y-axis variable and vice versa.

Inferential statistics

Particularly in research (e.g. in setting a hypothesis or question) and in some undergraduate experiments, we need to know whether the mean results differ from the controls significantly (i.e. the results did not occur by chance). There is a whole range of tests which can be used to calculate this. One of the most common is the *Student's t-test*. You may also need to learn *analysis of variance* (ANOVA) to do multiple comparisons of means. There are many statistical books which will guide you through these tests, including the use of software packages such as Microsoft *Excel, GraphPad Prism* and *SPSS*.

Appendix 6: Using Software to Draw Tables, Graphs and Calculating Descriptive Statistics

There is a wide variety of software to present data in tables and graphs and to analyse data statistically. The instructions which follow below to draw tables and graphs and to analyse data are using the Microsoft Office 2007 software. Please note that there may be differences depending on the version of the software being used.

Drawing a table using Microsoft Word (using MS Office 2007)

Step 1: Start Microsoft *Word*, then click the *Office* button, click 'New', then *Blank document* and then *Create*.

Step 2: Click on the 'Insert' tab, then click *Table*, then *Insert table*.

Step 3: Put in the number of columns and number of rows and click OK.

Drawing a table using Microsoft Excel

Step 1: Open Microsoft *Excel* to show a spreadsheet with columns labelled with the letters of the alphabet starting from 'a' and rows numbered from 1 onwards.

Step 2: Type the names of variables in the first row, with the units in the second row.

Step 3: Type the subject number or initials in the first column.

Step 4: Type in your data.

The table below shows the personal data and the heart rate at rest during a maths test in eleven subjects.

Essential Laboratory Skills for Biosciences, First Edition. M.S. Meah and E. Kebede-Westhead.
© 2012 John Wiley & Sons, Ltd. Published 2012 by John Wiley & Sons, Ltd.

APPENDIX 6: SOFTWARE FOR TABLES, GRAPHS AND STATISTICS

Table A6.1 *Example of a table using Microsoft* Word *or* Excel

Subject	Gender	Age (yrs)	Weight (kg)	Height (cm)	HR$_{rest}$ (b/min)	HR$_{maths}$ (b/min)
CA	M	19	74	182	67	69
NS	M	25	75	181	72	80
RE	M	25	97	186	88	98
JT	F	18	70	166	71	100
SO	M	21	75	177	80	90
SH	F	25	45	161	75	110
HC	F	19	65	170	78	98
OP	M	25	82	185	84	87
PJ	M	24	85	179	90	105
JH	F	19	58	173	64	70
KD	F	32	63	169	60	95

Using the data in Table A6.1, the descriptive statistics are calculated as follows:

Calculating descriptive statistics using Microsoft Excel

Step 1: Click on *Data*, then *Data analysis* (if you do not see data analysis, you will need to install it using 'Add-ins' from the *Home* button).

Step 2: Click on descriptive statistics, then OK.

Step 3: Click the option to choose whether your data is in columns or rows.

Step 4: In the input range, specify the data for which you want to calculate the descriptive stats (e.g. if your data was in column A, and in the first three rows, you would type 'A1:A3'). If you had more than one variable, then these could all be entered in one go by specifying the location. An easier method of inputting the data is to click the red arrow in the input box to go to the data; then highlight the required data and click the red arrow again, which will return you to the input box. The data location is now shown in terms of column and row labels. **Precaution**: check that these labels are correct.

Step 5: Choose where you want the results, either on the same sheet (in which case you have to input the cell number) or on a new worksheet.

Table A6.2 *Descriptive statistics*

	Age	Weight	Height	HR_{rest}	HR_{maths}
Mean	22.91	71.73	175.36	75.36	91.09
N	11	11	11	11	11
±SD	4.2	14.1	8.2	9.7	13.5
±SEM	1.3	4.2	2.5	2.9	4.1

Step 6: Tick the box showing summary statistics and press OK.
Step 7: You will then see 13 terms which represent the descriptive statistics for your data. Choose N, Mean, SD and SEM (see Table A6.2).

Using the mean and standard error of the mean (SEM) the mean changes in heart rate from rest to maths can be shown on a bar graph as follows (see Figure A6.1):

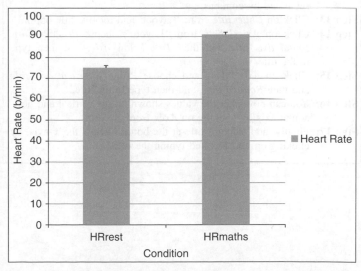

Figure A6.1 *Bar graph showing changes in the mean heart rate (±SEM) of students at rest and during a maths test*

Drawing a bar graph

Step 1: Open Microsoft *Excel* and a file showing your data.

Step 2: Click on *Insert*, then *Column*.

Step 3: Choose *Clustered column (2D)*; an empty rectangular box will appear.

Step 4: Choose *Select dat*a.

Step 5: Delete anything in the chart data range.

Step 6: Click *Add* (under 'Legend entries').

Step 7: In the *Series name* box, type the variable name, i.e. 'Heart rate'.

Step 8: In the *Series values* box, type the mean values.

Step 9: Click *Edit* in the horizontal axis labels and type in the labels for each bar.

Step 10: Now click on *Layout* from the top row and choose *Error bars*.

Step 11: Click *More error bar options* and then choose how to display the error bars, e.g. 'Plus side only'.

Step 12: Click *Custom* and then specify value and type in SEM values separated by commas.

Step 13: Click on *Chart title* from 'Layout' and remove title option.

Step 14: Click on *Axis labels* from 'Layout', choose *Primary horizontal axis title* and then *Title below axis*, and then type in the title.

Step 15: Click on *Axis labels* and choose *Primary vertical axis title* and then *Rotated title*, and then type in the title.

Step 16: You can now alter the values shown on the vertical and horizontal axis by clicking on *Axes* from 'Layout'.

Step 17: Finally, add a figure title at the bottom below the x-axis, by clicking on text box and typing the title.

Index

Essential Laboratory Skills for Biosciences, First Edition. M.S. Meah and E. Kebede-Westhead.
© 2012 John Wiley & Sons, Ltd. Published 2012 by John Wiley & Sons, Ltd.